Discrete-Time Higher Order Sliding Mode

Nalin Kumar Sharma
Janardhanan Sivaramakrishnan

Discrete-Time Higher Order Sliding Mode

The Concept and the Control

 Springer

Nalin Kumar Sharma
Department of Electrical Engineering
Indian Institute of Technology Delhi
New Delhi, India

Janardhanan Sivaramakrishnan
Department of Electrical Engineering
Indian Institute of Technology Delhi
New Delhi, India

ISBN 978-3-030-13088-6   ISBN 978-3-030-00172-8   (eBook)
https://doi.org/10.1007/978-3-030-00172-8

This Springer imprint is published by the registered company Springer Nature Switzerland AG
The registered company address is: Gewerbestrasse 11, 6330 Cham, Switzerland

The culture of research in India is very old and rich. Though the ancient sanskrit scriptures are full of knowledge, yet they kindle new thoughts to its readers and embrace new ideas.

ā no bhadrāh kratavo yantu viśvato adabdhāso aparītāsa udbhidah |

English Translation:

May noble thoughts come to us from every quarter, unchanged, unhindered, undefeated in every way;
  Rig Veda: 1.89.1

We want to dedicate this book to the Indian culture of research.

Preface

The aim of a control engineer is to control a specified system such that the desired performance is achieved. For this purpose, the system under consideration is modelled mathematically, but a disparity always lies between the actual system and its mathematical model. In these situations, robust control techniques provide schemes to control the system as per the requirement.

One of the major robust control technique is sliding mode control (SMC) technique, proposed by Emelyanov and coworkers in the 1970s. In this technique, a control input is designed such that the system's state trajectory reaches a prescribed manifold in a finite time and thereafter, remains on it despite the presence of uncertainties in the system. To establish and maintain the sliding mode, a control is designed such that the state trajectory is always directed towards the sliding manifold. To satisfy this constraint, SMC utilizes the concept of variable structure control (VSC). In VSC, the structure of the control law changes on the basis of the state information such that the system's state trajectory remains on the sliding manifold. Traditionally, a high-frequency switching control input is utilized to force the system dynamics to sliding mode despite the uncertainties in the system. However, the high-frequency switching in the control input provokes the problem of undesirable high-frequency vibration, termed as chattering, in the closed-loop system. The chattering effect can cause the wear and tear in mechanical system and is a major obstacle in the application of sliding mode technique.

Among the various approaches applied to combat chattering, the most promising is the concept of higher order sliding mode (HOSM) which was developed by A. Levant. The higher order sliding mode based control acts on the higher order time derivatives of sliding function rather than its first-order time derivative, as in the classical sliding mode, thus eliminating chattering without compromising robustness.

The utilization of computers and digital controllers for the implementation of control algorithms has been increasing in the recent times. In digital circuits, typically, the control input can be applied only at certain sampling instants and thereafter, remains constant for that entire sampling period. Thus, in computer controlled systems, the sliding mode control design for the discrete-time representation of a system is more justifiable, rather than its continuous-time model. As a result, the interest and research in the sliding mode control for discrete-time systems, aptly termed discrete-time sliding mode (DSM), have been remarkably increased. Generally, two approaches have been considered in literature to design discrete sliding mode. That is to design sliding mode control algorithms for discrete-time system representation, or discretization of the continuous-time sliding mode algorithms to obtain their discrete-time counterparts. However, discretization of a continuous sliding mode is generally unreliable and may eventually not retain the robustness and stability properties of sliding mode. Thus, it would be preferable to design the discrete-time counterpart of a continuous-time sliding mode control, rather than discretize a control designed based on continuous-time control concepts.

Motivation

The continuous-time higher order sliding mode improves the accuracy of sliding mode and smoothness of control input. Further, it eliminates the chattering problem in continuous-time sliding mode. The higher order sliding mode (HOSM) algorithms designed for continuous-time systems are generally applied through computer controlled circuits. Therefore, HOSM control algorithm designed for a discrete-time representation of a system is more justifiable. These facts motivate to develop the higher order sliding mode for discrete-time systems. However, in contrast to continuous-time sliding function and its continuous higher derivatives, discrete-time sliding function and its higher order differences are never continuous in nature. Hence, the definition and concept of continuous-time higher order sliding mode cannot be directly extended in discrete-time system representation. Therefore, there is also a need to conceptualize the discrete-time version of HOSM.

Organization

The prime objective of this book is to introduce and initiate the concept of *Discrete-Time Higher Order Sliding Mode* (DHOSM). To create a necessary base for the better understanding of the concept, a few fundamentals of sliding mode, higher order sliding mode and discrete-time sliding mode are briefed in Chap. 1. The rest of the work delves into the concept of DHOSM and its application to a variety of systems.

1. The basic concepts of sliding mode and sliding mode control are given in Chap. 1. This chapter also contains the brief, but necessary, knowledge of higher order sliding mode and discrete-time sliding mode for preparing the readers to understand the concepts presented in this book.

2. A definition of discrete-time higher order sliding mode (DHOSM) for a discrete-time system is explored in Chap. 2. Based on the definition of DHOSM, a generalized discrete-time reaching law is proposed to design a second-order sliding mode control for a discrete-time linear time-invariant (LTI) system. Subsequently, the behaviour and stability of the resulting second-order sliding mode are analysed.

3. The idea of an optimal discrete-time higher order sliding mode is explored in Chap. 3. A surface is designed for an LTI system such that when system's state trajectory is confined to it, a specified linear quadratic function is minimized over a fixed time period with fixed final state, and ensure DHOSM in spite of disturbance in the system. A disturbance estimation technique is used to reduce the DHOSM bandwidth further.

4. The disturbances affecting a system need not necessarily affect the system though the input channel, hence, discrete-time higher order sliding mode control for an LTI system in the presence of the unmatched uncertainties is studied in Chap. 4. A disturbance forecasting technique is used to forecast the disturbance to design DHOSM control.

5. In many cases, it is very difficult to estimate the bounds on the external disturbances affecting a system. Thus, it may be necessary to adapt the gains of the controller to the uncertainty level. In Chap. 5, a higher order reaching law with the adaptive switching gain is explored, which does not require the knowledge of the bound on the uncertainty and thus, enhances the practical applicability of the controller.

6. A system can also have probabilistic uncertainties. To accommodate such systems, the concept of discrete-time higher order sliding mode is extended to stochastic systems and a definition of stochastic discrete higher order sliding mode (SDHOSM), in probabilistic sense, is introduced in Chap. 6. An appropriate control law which steers uncertain stochastic system to the defined SDHOSM takes place in a band is also explored.

The aim of this work is to bridge the gap between continuous-time sliding mode and discrete-time sliding mode, by bringing in many concepts that are well defined in the former domain, into the latter domain. It is written in a manner that graduate students, who are interested in sliding mode control, and particularly, the discrete-time variety, would be able to grasp the difference in the design philosophy of continuous and discrete sliding mode, and we hope that it would pave a way for further future research in the area of application based discrete-time sliding mode control.

New Delhi, India   Nalin Kumar Sharma
                                                              Janardhanan Sivaramakrishnan

Acknowledgements

To our parents, teachers, and family, for the knowledge they imparted, and the support given to us, without which we would not have been capable of writing this book.

Contents

Acronyms

Abbreviations

ADHOSM	Adaptive discrete-time higher order sliding mode
DHOSM	Discrete-time higher order sliding mode
DSM	Discrete-time sliding mode
HOSM	Higher order sliding mode
LTI	Linear time-invariant
ODHOSM	Optimal discrete-time higher order sliding mode
SDHOSM	Stochastic discrete-time higher order sliding mode
SMC	Sliding mode control

Symbols

$\mathbb{R}$	Set of real numbers
Ω	Sample space
$\mathbb{E}\{.\}$	Expectation operator
$\mathbb{E}\{.|.\}$	Conditional expectation operator
A	State matrix in discrete-time model of LTI system
B	Input matrix in discrete-time model of LTI system
$\tilde{d}$	Disturbance vector in a system
d	The effect of the disturbance vector on the sliding function
d^*	bound of d
d_l	Lower bound of d
d_u	Upper bound of d
d_0	The mean of disturbance bounds d_u and d_l
d_x	The disturbance in state x
d_ξ	The disturbance in state ξ
I	Identity matrix
k	Sampling instant
P	Probability measure

k	Spring constant in N/m
m	Mass in kg
c	Damping coefficient in N/(m/s)
ε	Switching gain
ε^*	Upper bound of ε
n	Number of the states in an LTI system
m	Number of the inputs in an LTI system
r	Sliding order
$\bar{s}$	Higher order sliding function
k_1	The variable denoting the gain
k_2	The variable denoting the gain
T	Sampling time
u_{opt}	Optimal control
u_f	Feedback control
u_{eq}	Equivalent control
u_{sto}	Stochastic control
ϕ_i	Weight coefficient
ξ	The vector containing the sliding function and its higher order terms
V	Lyapunov function
V_w	Variance of signal w

Chapter 1
Preliminaries

1.1 Sliding Mode

Definition 1.1 *Sliding Mode*: [1] Sliding motion or sliding mode may be defined as the evolution of the state trajectory of a system confined to a specific non-trivial submanifold of the state space with stable dynamics. The said submanifold is termed as sliding manifold or sliding surface.

Since the closed-loop system's state trajectory remains on the sliding manifold in the sliding mode, the sliding manifold governs the dynamics of the closed-loop system's state trajectory. Therefore, the sliding manifold is designed such that the desired closed-loop performance is achieved [2].

To establish and maintain the sliding mode, the system's state trajectory must converge to the sliding manifold from all the directions of the state space, i.e. the sliding manifold must have attractive nature [3]. If a function $s(x, t)$ is defined as the sliding function, then $s(x, t) = 0$ represents the sliding manifold. The attractiveness of the sliding manifold can be ensured mathematically by

$$\lim_{s(x,t) \to 0^+} \dot{s}(x, t) < 0 \quad \text{and} \quad \lim_{s(x,t) \to 0^-} \dot{s}(x, t) > 0,$$

or in the composed form

$$s(x, t)\dot{s}(x, t) < 0, \tag{1.1}$$

in some domain $\Omega \subset \Re^n$ [2]. The condition (1.1) is known as the reaching condition [4]. A more stronger condition of reachability is the η-reaching condition [3]. It ensures the establishment of sliding mode in finite time and expressed as

$$s(x, t)\dot{s}(x, t) < \eta |s(x, t)|, \eta > 0. \tag{1.2}$$

© Springer Nature Switzerland AG 2019
N. K. Sharma and J. Sivaramakrishnan, *Discrete-Time Higher Order Sliding Mode*,
https://doi.org/10.1007/978-3-030-00172-8_1

As a result in the sliding mode, the sliding manifold becomes an invariant set in the finite time. Let us consider a dynamical system

$$\dot{x} = f(x, t) + g(x, t)u(t), \tag{1.3}$$

where $f \in \mathbb{R}^{(n \times 1)}$ and $g \in \mathbb{R}^{(n \times 1)}$ are the continuous function of x and t; and $u \in \mathbb{R}$ is the control input. Consider a sliding function

$$s(x, t) \in \mathbb{R}. \tag{1.4}$$

The derivative of s is found out as

$$\dot{s} = \frac{\partial s}{\partial x} (f(x, t) + g(x, t)u(t)). \tag{1.5}$$

Intuitively, it is clear that $\dot{s}$ has to be discontinuous to fulfil the reaching condition (1.1). Since f and g are continuous function, the control input may need to be discontinuous and in the form

$$u(x, t) = \begin{cases} u^+(x, t) & \text{if } s(x, t) > 0 \\ u^-(x, t) & \text{if } s(x, t) < 0 \end{cases} \tag{1.6}$$

The discontinuous control (1.6) causes two different state velocities on either side of the sliding manifold represented as

$$\dot{x} = \begin{cases} f^+(x, t) = f(x, t) + g(x, t)u^+(x, t) & \text{if } s(x, t) > 0 \\ f^-(x, t) = f(x, t) + g(x, t)u^-(x, t) & \text{if } s(x, t) < 0 \end{cases} \tag{1.7}$$

The control inputs $u^+(x, t)$ and $u^-(x, t)$ in (1.7) are designed such that the state velocities $f^+(x, t)$ and $f^-(x, t)$ are directed towards the sliding manifold. As a result, all the state trajectories converge on the sliding manifold and make it an attractive invariant set.

Equation (1.7) is a differential equation with discontinuous right-hand side. It can be noticed that the solution of (1.7) exists either for the region $s(x, t) > 0$ or $s(x, t) < 0$, but the existence and uniqueness of solution of (1.7) on the sliding manifold $s = 0$ is not very clearly defined in the classical sense. Since the sliding mode takes place on the sliding manifold, the existence and uniqueness of the solution of (1.7) on the sliding manifold is fundamentally crucial. However, the equations in (1.7) do not provide a straightforward indication regarding the velocity vector on the sliding manifold.

Filippov addressed this issue and provided a method to solve the differential equations with discontinuous right-hand side [5]. It is demonstrated that the solution of (1.7) with the control (1.6) onto the discontinuity manifold satisfies the differential inclusion

$$\dot{x} \in V(x, t) \tag{1.8}$$

where $V(x, t)$ is the minimal convex closure containing all values of $f(x, t, u(x, t))$ when x covers a δ-neighbourhood (where $\delta \to 0$) of the manifold. It is stated that the resulting velocity vector of the state trajectory of (1.7) on the sliding manifold $s(x, t) = 0$ is the solution of

$$\dot{x} = \alpha f^+(x, t) + (1 - \alpha)f^-(x, t) = f^0(x, t), \tag{1.9}$$

where $\alpha \in [0, 1]$ is chosen such that the vector $f^0(x, t)$ lies in the plane tangential to the sliding manifold. The end of vector $f^0(x, t)$ is the intersecting point of the tangential plane and the line connecting the end of vectors f^+ and f^- as shown in Fig. 1.1. Thus, the solution of (1.7) with the control (1.6) exists and uniquely defined on the sliding manifold $s(x, t) = 0$ as

$$\dot{x} = \frac{\frac{\partial s}{\partial x} f^-}{\frac{\partial s}{\partial x}(f^- - f^+)} f^+ - \frac{\frac{\partial s}{\partial x} f^+}{\frac{\partial s}{\partial x}(f^- - f^+)} f^-. \tag{1.10}$$

This solution is known as the solution in the Filippov sense [1].

Another approach to determine the state velocity on the sliding manifold is the equivalent control method introduced by Utkin [3]. The equivalent control provides solution identical to the solution obtained through Filippov method, however its simpler analysis and design procedure have encouraged its widespread utilization in the literature. The equivalent control drives the system trajectory on the sliding manifold if the system trajectory is already on the sliding manifold, i.e. if the sliding mode is established already. Mathematically, in the presence of the equivalent control input u_{eq}, the system trajectory satisfies $s(x, t) = 0$ as well as $\dot{s}(x, t) = 0$, i.e.

$$\dot{s} = \frac{\partial s}{\partial t} + \frac{\partial s}{\partial x} f(x, t) + \frac{\partial s}{\partial x} g(x, t)u_{eq}(x, t) = 0. \tag{1.11}$$

The equivalent control can be computed from the above equality as

Fig. 1.1 Illustration of velocity vector in Filippov method

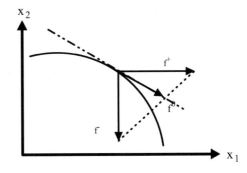

$$u_{eq}(x, t) = - \left(\frac{\partial s}{\partial x} g(x, t) \right)^{-1} \left(\frac{\partial s}{\partial t} + \frac{\partial s}{\partial x} f(x, t) \right). \tag{1.12}$$

Thus, under the action of the equivalent control, any system trajectory starting from the sliding manifold remains on the sliding manifold. Consider a time t_s such that $s(x, t_s) = 0$. The system dynamics on the sliding manifold for all $t \geq t_s$ can be obtained as

$$\dot{x} = \left[f(x, t) - g(x, t) \left(\frac{\partial s}{\partial x} g(x, t) \right)^{-1} \left(\frac{\partial s}{\partial t} + \frac{\partial s}{\partial x} f(x, t) \right) \right]. \tag{1.13}$$

The closed-loop dynamics (1.13) with the constraint $s(x, t) = 0$ governs the behaviour of the system on the sliding manifold. Therefore, the system dynamics must satisfy the n-dimensional state dynamics (1.13) as well as the m-dimensional constraint $s(x, t) = 0$. The utilization of both of these relations reduces the system's n-dimensional model to $(n - m)$-dimensional model. In this way, the sliding mode is governed by a reduced order dynamics.

It must be noted that the equivalent control is operative only when the system's state is on the sliding manifold. Therefore, an algorithm is required such that the sliding function reaches zero from any arbitrary initial condition. For this purpose, the reaching law method is introduced such that the dynamics of the sliding function is directly dictated through a specific reaching law. Various reaching laws have been proposed in the literature of the sliding mode [1].

The sliding motion involves two phases as depicted in Fig. 1.2. The first phase is called reaching phase, in which the trajectory starts from anywhere on the phase plane and moves towards the sliding manifold in the finite time. The second part is the sliding phase in which the trajectory asymptotically tends to the origin of the phase plane [6]. Once the sliding function reaches the sliding manifold, the reaching law starts behaving as the system in the effect of the equivalent control.

However, it should be noted that the equivalent control input requires perfect knowledge of the system parameters and no disturbances in the system to perfectly slide on the sliding manifold. This situation is practically impossible and thus, the

Fig. 1.2 The reaching and sliding phase in sliding mode

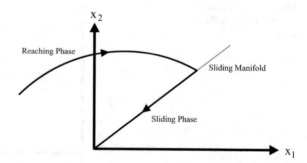

system under the effect of the equivalent control is not robust against the uncertainties. Nevertheless, the robustness property is ensured by adding a discontinuous control in the equivalent control. The discontinuous control part has two functions:

1. During the reaching phase, it ensures that the sliding function, starting from any initial condition other than the sliding manifold, reaches the sliding manifold in finite time.
2. During the sliding phase, if the sliding function leaves the sliding manifold, discontinuous control brings it back on the sliding manifold. Thus, it ensures that the sliding function remains on the sliding manifold in spite of the uncertainties.

Therefore, to retain the robustness property in the sliding mode, the discontinuous control is also required in addition with the equivalent control. We can understand the discontinuous control as a fast and brutal force which acts on the system even due to the minor deviation of the system trajectory from the sliding manifold. This conduct causes a high-frequency switching in the sliding function which is considered as the inherent property of the sliding mode. However, it results in undesired high-frequency vibration in the closed-loop system. This phenomenon is known as 'chattering' which causes wear and tear in the hardware [7–10]. Therefore, the chattering problem is the biggest drawback of the sliding mode technique which make this technique not suitable for the mechanical systems. Nevertheless, various techniques have been suggested to prevent this chattering problem [11].

One of these techniques is known as the boundary layer solution [4, 12]. In this technique, the discontinuity generated by the 'sign' function is replaced by the 'saturation function' in the vicinity of sliding manifold to eliminate the chattering. However, this smooth approximation technique results in deterioration of the accuracy and robustness of the sliding mode.

Another approach to eliminate the chattering is the higher order sliding mode control. Since the focus of the book is on the higher order sliding mode, this technique is described hereby in detail.

1.2 Higher Order Sliding Mode

The main idea of the higher order sliding mode (HOSM) is that the control input forces the sliding function and its first $(r - 1)$ higher order derivatives to zero by acting discontinuously on the r-th time derivative of the sliding function. Thus, it keeps the intrinsic nature of the sliding mode by shifting the discontinuity on the r-th derivative of the sliding function.

Definition 1.2 *Sliding Set*: [13] Let a constraint be given by a function $s = 0$, where $s : R^n \rightarrow R$ is a sufficiently smooth function. It is also supposed that total time derivatives along the trajectories $s, \dot{s}, \ddot{s}, \ldots, s^{r-1}$ exist, i.e. discontinuity does not appear in the first $(r - 1)$ total time derivatives of the sliding function s. Then, the r-th order sliding set is determined by the equalities

$$s = \dot{s} = \ddot{s} = \cdots = s^{(r-1)} = 0. \tag{1.14}$$

The integer r is called sliding order.

Note that the sliding set, defined by (1.14), imposes an r-dimensional condition on the state of the dynamic system.

Definition 1.3 *r-sliding mode*[13] Let the r-sliding set (1.14) be non-empty and assume that it is locally an integral set in Filippov's sense (i.e. it consists of Filippov's trajectories of the discontinuous dynamic system). Then, the corresponding motion satisfying (1.14) is called an r-sliding mode with respect to the function s.

The higher order sliding mode can be further classified into ideal HOSM and real HOSM.

Definition 1.4 *Ideal Higher Order Sliding Mode* [13] Ideal sliding takes place when the state trajectory perfectly slides along the r-dimensional manifold defined by the sliding set (1.14). The sliding manifold generated by (1.14) is called ideal higher order sliding manifold.

As it is clear from the definition that the ideal higher order sliding occurs when the sliding function perfectly remains on the sliding manifold. This ideal situation can be realized only if there are no switching imperfections and the switching frequency is infinite, which is impossible in practice. Therefore, a more practical sliding mode is introduced, which is known as real sliding mode. In real sliding, the switching imperfections are taken into account and the sliding function is kept only at the vicinity of the zero. Let ε be some measure of the switching imperfections. This imperfect switching may be due to non-zero switching interval, delay, measurement interval, switching discretization or the other irregularities.

Definition 1.5 *Real Higher Order Sliding Mode* [14] Let $\gamma(\varepsilon)$ be a real-valued function such that $\gamma(\varepsilon) \to 0$ as $\varepsilon \to 0$. A real sliding algorithm on the constraint $s = 0$ is said to be of order r $(r > 0)$ with respect to $\gamma(\varepsilon)$ if for any compact set of initial conditions and for any time interval $[T_1, T_2]$, there exists a constant C, such that the steady-state process for $[T_1, T_2]$ satisfies

$$s(t, x(t, \varepsilon)) \leq C|\gamma(\varepsilon)|^r. \tag{1.15}$$

The above definition states that if a system is in r-sliding mode with minimal switching interval τ, the corresponding sliding precision is $|\tau|^r$. Thus, r-sliding mode realization can provide for up to r-th order of sliding precision with respect to the switching imperfection.

Proposition 1.1 [14] *Let l be a positive number not exceeding r. If l-th derivative of s is bounded in ε for some steady part of $x(t, \varepsilon)$, then there exists positive constants $C_1, C_2, \ldots, C_l$ such that for the steady-state process, the following inequalities hold:*

$$\|\dot{s}\| \leq C_1 \tau^{l-1}, \|\ddot{s}\| \leq C_2 \tau^{l-2}, \ldots, \left\|s^{(l-1)}\right\| \leq C_{l-1}\tau. \tag{1.16}$$

This proposition asserts that the derivatives of the sliding function of a system with r-sliding mode exhibit certain accuracy in the steady state. If the minimal switching interval is τ, then in the r-sliding mode, the corresponding sliding precision of the derivatives of the sliding function is always satisfied by inequality

$$\|s^{(i)}\| \leq C_i |\tau|^{r-i}, \qquad \forall i = 0 \quad \text{to} \quad r - 1. \tag{1.17}$$

Let us illustrate the design of a higher order sliding mode with sliding order $r = 2$. A system defined in (1.3) as

$$\dot{x} = f(x, t) + g(x, t)u(t),$$

with a sliding function $s(x, t)$. The objective is the stabilization of s and $\dot{s}$ function at the zero. The second-order sliding mode is designed on the basis of the relative degree of the system.

Definition 1.6 *Relative degree*: [15, p. 139] The relative degree is defined as the number of times one has to differentiate the output variable in order to have a direct relation between the output variable and the control input.

In our case, the sliding function can be understood as an output variable. Therefore, the relative degree is the total number of derivatives of the sliding function required to get a direct relation with the control input. Mathematically, the sliding function has relative degree i, if $\frac{\partial s}{\partial u} = \frac{\partial \dot{s}}{\partial u} = \cdots = \frac{\partial s^{(i-1)}}{\partial u} = 0$ and $\frac{\partial s^{(i)}}{\partial u} \neq 0$. On using the concept of the relative degree, there are two possible cases in second-order sliding mode:

Case A: *Relative degree one*

The first derivative of s is

$$\dot{s}(x, t) = \alpha_1(x, t) + \beta_1(x, t)u(t), \tag{1.18}$$

where

$$\alpha_1(x, t) = \frac{\partial s(x, t)}{\partial t} + \frac{\partial s(x, t)}{\partial x} f(x, t),$$

and

$$\beta_1 = \frac{\partial s(x, t)}{\partial x} g(x, t).$$

In this case, the relative degree is considered to be one. Thus, $\beta_1 \neq 0$, i.e. there is a direct relation between $\dot{s}$ and the control input u. For this case, the second derivative of the sliding function can be obtained as

$$\ddot{s}(x, t) = \alpha_{2A}(x, t) + \beta_1(x, t)\dot{u} \tag{1.19}$$

where

$$\alpha_{2A}(x, t) = \frac{\partial \alpha_1(x, t)}{\partial t} + \frac{\partial \alpha_1(x, t)}{\partial x} f(x, t) + \left(\frac{\partial \alpha_1(x, t)}{\partial x} g(x, t) + \frac{\partial \beta_1(x, t)}{\partial t} + \frac{\partial \beta_1(x, t)}{\partial x} \dot{x} \right) u(x, t)$$

is known as the drift term. The control input u is considered as an unknown disturbance affecting $\alpha_{2A}(x, t)$. The derivative of control input, $\dot{u}$, is considered as the auxiliary control input and it is designed such that it steers s as well as $\dot{s}$ to zero. To retain the robustness property of sliding mode, $\dot{u}$ is designed such that it has discontinuous nature. Once the auxiliary control input, $\dot{u}$, is obtained, the actual control, u, can be found out by integrating $\dot{u}$. Thus, even though $\dot{u}$ is a discontinuous control, after the integration, u is obtained as a continuous control input and problem of chattering is eliminated. Therefore, this case is also known as '**Antichattering case**' [16].

Case B: *Relative degree two*

In this case, the relative degree is considered to be two, so $\beta_1 = 0$ and the control input u has a direct relation with the sliding function $\ddot{s}$. The first derivative of the sliding function is

$$\dot{s}(x, t) = \alpha_1(x, t), \tag{1.20}$$

and the second derivative is

$$\ddot{s}(x, t) = \alpha_{2B}(x, t) + \beta_2(x, t)\dot{u}, \tag{1.21}$$

where

$$\alpha_{2B} = \frac{\partial \alpha_1(x, t)}{\partial t} + \frac{\partial \alpha_1(x, t)}{\partial x} f(x, t)$$

and

$$\beta_2(x, t) = \frac{\partial \alpha_1(x, t)}{\partial x} g(x, t). \tag{1.22}$$

Here, the control input is computed such that the control signal affects $\ddot{s}$ using s and $\dot{s}$ or using s and sgn$[\dot{s}]$. Various second-order sliding mode control algorithms have been reported in the literature since the inception of the higher order sliding mode. Some of the major algorithms are twisting controller [14], suboptimal algorithm [17–21], super-twisting control [14], and drift algorithm [13, 14]. The other major works in the higher order sliding mode control design can be found out in [22–31].

Another major point is the finite-time convergence of the higher order sliding mode. While it is easier to inspect the finite-time reachability property and to find the settling time in the first-order sliding mode, the finite-time convergence of s and its higher order derivatives in HOSM is a non-trivial task [32]. In the literature of higher order sliding mode, it was shown that the convergence to higher order sliding mode may be asymptotic as well [13]. However, the finite-time stability of the higher order sliding mode and computation of the settling time are rigorously studied in the last decade. In particular, the property of finite-time stability of the super-twisting algorithm has attracted the attention of the control community and as a result, significant improvements are made in this algorithm [33–36].

1.3 Discrete-Time Sliding Mode

With the advent of the digital computers and microprocessor, nowadays, the control strategies are applied through the digital circuitry. This modernization encouraged the development of control strategies for discrete-time systems. Unlike continuous-time control, digital control can be applied only at the sampling instants and remains constant, or 'freeze', throughout the sampling period. However, from the discussion of the concept of sliding mode, it is clear that the sliding mode requires infinite switching frequency to deliver its requisite properties. Therefore, due to the limited switching frequency constraint, the exact sliding mode in discrete-time system is impossible. The finite switching frequency leads to zig-zag motion of the sliding function in the vicinity of sliding manifold, and this way sliding mode in discrete-time system loses its invariance property. Thus, discrete nature of sliding mode in discrete-time systems demands a more realistic definition and design procedure.

Milosavljevic [37] gave an idea of discrete-time sliding mode (DSM) using quasi-sliding mode, where quasi-sliding mode is defined as the oscillation of the discrete-time sliding function in a small neighbourhood of the sliding manifold. A similar concept is provided and termed as pseudo-sliding mode in [38–40]. The sliding condition to ensure quasi-sliding mode, inspired by the reaching condition (1.1) of continuous-time sliding mode, is provided as

$$\lim_{s(k)) \to 0^+} [s(k+1) - s(k)] < 0, \quad \lim_{s(k)) \to 0^-} [s(k+1) - s(k)] > 0. \tag{1.23}$$

Fig. 1.3 The reaching and sliding phase in discrete-time sliding mode

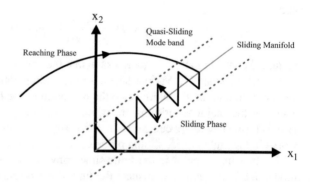

Similarly, Dote and Hoft [41] used an equivalent form of the continuous reaching condition to give a discrete reaching condition as

$$[s(k+1) - s(k)]s(k) < 0. \tag{1.24}$$

However, the reaching laws (1.23) and (1.24) do not impose a sufficient condition for the convergence of sliding function on the sliding manifold. Sarpturk et al. proposed a necessary and sufficient condition of the discrete-time sliding mode in [42]. This condition is stated as

$$|s(k+1)| \leq |s(k)|. \tag{1.25}$$

Hence, a control input synthesized through the reaching law (1.25) guarantees that the state trajectories remain in a decreasing or, at the worst, a non-increasing band around the sliding manifold as depicted in Fig. 1.3. This band is called as quasi-sliding mode band. Further, Drakunov et al. [43] defined the discrete-time sliding mode as

Definition 1.7 *Discrete-time sliding mode*: [43] We say that in the discrete-time dynamical system $x(k+1) = F[x(k)]$, $x(k) \in \mathbb{R}^n$, a discrete-time sliding mode takes place on the subset M of a manifold

$$\sigma = \{x : s = 0\}, \tag{1.26}$$

if there exists an open neighbourhood U of this subset such that from $x \in U$, it follows $F(x) \in M$.

Sliding mode control community has two major schools of thought regarding the discrete-time sliding mode control design. The first is to discretize an available continuous-time sliding mode control technique to obtain its discrete-time counterpart. Various first-order sliding mode techniques [44–47] and higher order sliding mode techniques [48–51] are obtained using discretization. However, discretization of a continuous-time sliding mode algorithm may lead to periodic behaviour [52–54], or at worst, unstable closed-loop system [55, 56].

In the second framework, a discrete-time sliding mode control algorithm is designed for a discrete-time system explicitly. Following this framework, the reaching law (1.25) has inspired two classes of reaching laws:

1. Switching type of reaching laws [6, 57–71]
2. Non-switching type of reaching laws [1, 47, 55, 72–80]

Gao et al. [6] presented a switching type of the reaching law and three general conditions for DSM, which were

1. Starting from any initial state, the trajectory will move monotonically toward the switching plane (sliding manifold) and cross it in finite time.
2. Once the trajectory has crossed the switching plane the first time, it will cross the plain again in every successive sampling period, resulting in a zigzag motion about the switching plane.
3. The size of each successive zig-zagging step is non-increasing and the trajectory stays within a specified band.

On the basis of these three conditions, Gao et al. defined quasi-sliding mode as:

Definition 1.8 *Quasi-sliding mode*: [81] The motion of a discrete VSC system satisfying condition-1 and condition-2 is called a quasi-sliding mode (QSM). The specified band which contains the QSM is called the quasi-sliding mode band (QSMB) and is defined by $x| - \triangle < s(x, t) < +\triangle$, where $2\triangle$ the width of the band.

Following the definition of the quasi-sliding mode, a reaching law is proposed in [6], which is further modified in [82] as

$$s(k + 1) = (1 - qT)s(k) - \varepsilon T \operatorname{sgn}(s(k)), \qquad (1.27)$$

where q and ε are the constant parameters and T is the sampling time. The parameters are chosen such that $(1 - qT) > 0, q > 0$ and $\varepsilon > 0$.

Later, Bartoszewicz [73] eased the general conditions for DSM given by Gao [6] and proposed a non-switching type of reaching law. The sliding function need not to cross the sliding manifold at every successive sample but to always remain in a small band around it. The non-switching type of reaching law is based on the concept that a discrete-time sliding mode control law need not necessarily be of variable structure or need not to have explicit discontinuity as in (1.27). The non-switching type of reaching laws tries to satisfy the condition $s(k + 1) = 0$ and apparently, the control is now no longer of variable structure.

Due to the widespread applicability of the discrete-time systems, the higher order sliding mode in discrete-time system has attracted the attention of the research community. Following this direction, some techniques are attempted to design second-order sliding mode by extending the concept of discrete-time first-order sliding mode to second-order [83–86]. However, as per the authors' knowledge, HOSM for discrete-time system has been never defined explicitly. Although, a few continuous-time higher order sliding mode algorithms are also discretized to obtain its discrete counterpart [48, 53, 87].

References

1. B. Bandyopadhyay, S. Janardhanan, *Discrete-Time Sliding Mode Control : A Multirate-Output Feedback Approach* (Ser. Lecture Notes in Control and Information Sciences, vol. 323, Springer, 2005)
2. C. Edwards, S. Spurgeon, *Sliding Mode Control: Theory and Applications* (CRC Press, 1998)
3. V. Utkin, IEEE Trans. Autom. Control **22**(2), 212 (1995)
4. V. Utkin, J. Guldner, J. Shi, *Sliding Mode Control in Electro-Mechanical Systems, Second Edition, Ed. Automation and Control Engineering* (CRC Press, 2009)
5. A. Filippov, *Differential Equations with Discontinuous Right hand Sides* (Springer, Netherlands, 1988)
6. W. Gao, Y. Wang, A. Homaifa, IEEE Trans. Ind. Electr. **42**(2), 117 (1995)
7. A. Levant, IEEE Trans. Autom. Control **55**(6), 1380 (2010)
8. A. Rosales, Y. Shtessel, L. Fridman, C.B. Panathula, IEEE Trans. Autom. Control **PP**(99) (2016)
9. I. Boiko, L. Fridman, IEEE Trans. Autom. Control **50**(9), 1442 (2005)
10. G. Bartolini, Int. J. Syst. Sci. **20**(12), 2471 (1989)
11. H. Lee, V.I. Utkin, Ann. Rev. Control **31**(2), 179 (2007)
12. J.J.E. Slotine, Int. J. Control **40**(2), 421 (1984)
13. L. Fridman, A. Levant, *Higher-Order Sliding Mode* (CRC Press, 2002), pp. 53–102
14. A. Levant, Int. J. Control **58**(6), 1247 (1993)
15. A. Isidori, *Nonlinear Control Systems*, 3rd edn. (Springer, London, 1995)
16. A. Pisano, Second order sliding modes: Theory and applications. Ph.D. thesis, University of Cagliari, Italy (2000)
17. G. Bartolini, A. Ferrara, E. Usai, Int. J. Robust Nonlinear Control **7**(4), 299 (1997)
18. G. Bartolini, A. Pisano, E. Punta, E. Usai, Int. J. Control **76**(9), 875 (2003)
19. G. Bartolini, A. Ferrara, E. Usai, IEEE Trans. Autom. Control **43**(2), 241 (1998). https://doi.org/10.1109/9.661074
20. A. Ferrara, M. Rubagotti, IEEE Trans. Autom. Control **54**(5), 1082 (2009)
21. A. Ferrara, G.P. Incremona, IEEE Trans. Control Syst. Technol. **23**(6), 2316 (2015)
22. S. Laghrouche, F. Plestan, A. Glumineau, Automatica **43**(3), 531 (2007)
23. S. Laghrouche, F. Plestan, A. Glumineau, in *European Control Conference (ECC)* (Cambridge, U.K., 2003)
24. S. Laghrouche, F. Plestan, A. Glumineau, *Practical Higher Order Sliding Mode Control: An Optimal Control Based Approach with Application to Electromechanical Systems* (Springer, Berlin, Heidelberg, 2006), pp. 169–191
25. J. Davila, L. Fridman, A. Levant, IEEE Trans. Autom. Control **50**(11), 1785 (2005)
26. A. Estrada, L. Fridman, Automatica **46**(11), 1916 (2010)
27. A.F. de Loza, F.J. Bejarano, L. Fridman, Int. J. Robust Nonlinear Control **23**(7), 754 (2013)
28. A. Levant, Int. J. Control **76**(9–10), 924 (2003)

29. F. Dinuzzo, A. Ferrara, IEEE Trans. Autom. Control **54**(9), 2126 (2009)
30. I. Castillo, M. Jimnez-Lizrraga, E. Ibarra, J. Frankl. Inst. **352**(7), 2810 (2015)
31. G. Bartolini, A. Levant, A. Pisano, E. Usai, Int. J. Control **89**(9), 1747 (2016)
32. V. Utkin, in *2010 11th International Workshop on Variable Structure Systems (VSS)* (2010), pp. 528–533
33. J.A. Moreno, M. Osorio, IEEE Trans. Autom. Control **57**(4), 1035 (2012)
34. A. Polyakov, A. Poznyak, IEEE Trans. Autom. Control **54**(8), 1951 (2009)
35. I. Nagesh, C. Edwards, Automatica **50**(3), 984 (2014)
36. M. Basin, C.B. Panathula, Y. Shtessel, IET Control Theor. Appl. **11**(8), 1104 (2017)
37. C. Miloslavjevic, Autom. Remote Control **46**, 679 (1985)
38. R.B. Potts, X. Yu, The journal of the australian mathematical society Series B. Appl. Math. **32**(4), 365 (1991). https://doi.org/10.1017/S0334270000008481
39. X. Yu, R.B. Potts, the journal of the australian mathematical society Series B. Appl. Math. **34**(1), 117 (1992). https://doi.org/10.1017/S0334270000007335
40. R. Potts, X.H. Yu, Comput. Math. Appl. **28**(1), 281 (1994)
41. Y. Dote, R.G. Hoft, in *Proceedings of the Industrial Applications Society Annual Meeting* (Cincinnati, OH, 1980)
42. S.Z. Sarpturk, Y. Istefabopulos, O. Kaynak, IEEE Trans. Autom. Control **32**(10), 930 (1987)
43. S.V. Drakunov, V. Utkin, in *IFAC Symp. Nonlinear Control Systems Design* (Italy, 1989)
44. X. Yu, B. Wang, Z. Galias, G. Chen, IEEE Trans. Autom. Control **53**(6), 1563 (2008)
45. B. Wang, X. Yu, G. Chen, Automatica **45**(1), 118 (2009)
46. S. Li, H. Du, X. Yu, IEEE Trans. Autom. Control **59**(2), 546 (2014)
47. A. Bartoszewicz, P. Leniewski, IEEE Trans. Control Syst. Technol. **24**(2), 670 (2016)
48. B. Wang, X. Yu, X. Li, IEEE Trans. Ind. Electr. **55**(11), 4055 (2008)
49. O. Huber, B. Brogliato, V. Acary, A. Boubakir, P. Franck, W. Bin, *Experimental results on implicit and explicit time-discretization of equivalent-control-based sliding-mode control* (IET, 2016)
50. V. Acary, B. Brogliato, Y.V. Orlov, IEEE Trans. Autom. Control **57**(5), 1087 (2012)
51. Y. Yan, Z. Galias, X. Yu, C. Sun, Automatica **68**, 203 (2016)
52. X. Xia, A. Zinober, Automatica **42**, 771 (2006)
53. A. Levant, D.L. Mili Livne, *On discretization of high-order sliding modes* (IET, 2016)
54. A. Levant, IFAC Proc. Vol. **44**(1), 1904 (2011)
55. S. Janardhanan, B. Bandyopadhyay, IEEE Trans. Autom. Control **51**(9), 1532 (2006)
56. A.K. Behera, B. Bandyopadhyay, Automatica **54**, 176 (2015)
57. G. Monsees, J.M.A. Scherpen, Int. J. Control **75**(4), 242 (2002)
58. S. Qu, X. Xia, J. Zhang, IEEE Trans. Ind. Electr. **61**(7), 3502 (2014)
59. D. Munoz, D. Sbarbaro, IEEE Trans. Ind. Electr. **47**(3), 574 (2000)
60. K. Abidi, J.X. Xu, Y. Xinghuo, IEEE Trans. Autom. Control **52**(4), 709 (2007)
61. Z. Xi, T. Hesketh, IET Control Theor. Appl. **4**(5), 889 (2010)
62. Z. Xi, T. Hesketh, IET Control Theor. Appl. **4**(10), 2071 (2010)
63. A. Bartoszewicz, P. Latosiski, Int. J. Robust Nonlinear Control **26**(1), 47 (2016)
64. Q. Xu, IEEE Trans. Robotics **29**(3), 663 (2013)
65. Q. Xu, IEEE Trans. Ind. Electr. **63**(6), 3976 (2016)
66. Q. Xu, IEEE Trans. Ind. Electr. **62**(12), 7738 (2015)
67. Q. Xu, IEEE Trans. Autom. Sci. Eng. **14**(1), 238 (2017)
68. O. Huber, V. Acary, B. Brogliato, IEEE Trans. Autom. Control **61**(10), 3016 (2016)
69. H. Du, X. Yu, M.Z. Chen, S. Li, Automatica **68**, 87 (2016)
70. S. Chakrabarty, B. Bandyopadhyay, Automatica **52**, 83 (2015)
71. S. Chakrabarty, B. Bandyopadhyay, Automatica **63**, 34 (2016)
72. S.K. Spurgeon, Int. J. Control **55**(2), 445 (1992)
73. A. Bartoszewicz, IEEE Trans. Ind. Electr. **45**(4), 633 (1998)
74. G. Golo, C. Milosavljevic, Syst. Control Lett. **41**(1), 19 (2000)
75. P. Ignaciuk, A. Bartoszewicz, IEEE Trans. Ind. Electr. **55**(11), 4013 (2008)
76. S. Hui, S.H. Ak. Syst. Control Lett. **38**(4), 283 (1999)

77. P. Ignaciuk, A. Bartoszewicz, IEEE Trans. Control Syste. Technol. **20**(5), 1400 (2012)
78. S. Janardhanan, B. Bandyopadhyay, IEEE Trans. Autom. Control **51**(6), 1030 (2006)
79. S. Janardhanan, B. Bandyopadhyay, IEEE Trans. Ind. Electr. **53**(5), 1677 (2006)
80. G. Bartolini, A. Ferrara, V.I. Utkin, Automatica **31**(5), 769 (1995)
81. W. Gao, J. Hung, IEEE Trans. Ind. Electr. **40**(1), 45 (1993)
82. A. Bartoszewicz, IETE J. Res. **43**(1), 235 (1996)
83. M. Mihoub, A.S. Nouri, R.B. Abdennour, Control Eng. Pract. **17**(9), 1089 (2009)
84. H. Romdhane, D. Khadija, N. Ahmed said. Int. J. Model. Identif. Control **22**, 159 (2014)
85. H. Romdhane, K. Dehri, A.S. Nouri, Int. J. Robust Nonlinear Control **26**(17), 3806 (2016)
86. A. Bartoszewicz, P. Latosiski, Int. J. Control **90** (2017)
87. I. Salgado, S. Kamal, B. Bandyopadhyay, I. Chairez, L. Fridman, ISA Trans. **64**, 47 (2016)

Chapter 2
Discrete-Time Higher Order Sliding Mode

2.1 Introduction

Discrete-time signal representations are inherently discontinuous and hence, the conventional definition of continuity and smoothness cannot be applied to them directly. Similarly, in contrast with continuous-time sliding function s and its continuous higher derivatives, discrete-time sliding function $s(k)$ and its higher order differences are always discrete in nature. Hence, the equivalent of higher order sliding mode for discrete-time systems is not clearly defined. Furthermore, the existing control algorithms are either designed for discrete-time first-order sliding mode [1–7] or discretized form of continuous-time second-order sliding mode [8–10].

The main objectives of this chapter are to explicitly define the concept of higher order sliding mode in discrete-time dynamical systems and to design a control law to demonstrate the applicability of the proposed concept. To achieve the above objectives, the following contributions are made in this chapter.

1. A definition of discrete-time higher order sliding mode (DHOSM) is proposed.
2. Based on the definition of DHOSM, a generalized discrete-time reaching law is proposed and utilized to design a second-order sliding mode control for an uncertain discrete-time LTI system. Further, the behaviour and the stability of the resulting second-order sliding mode are analysed.
3. The proposed algorithm is numerically simulated and experimentally validated on a real-time rectilinear plant.
4. To evaluate the performance of the proposed technique against other existing algorithms, two performance parameters, sensitivity function and (p, q_r)-continuity are considered.

This chapter is organized as follows: A definition of discrete-time higher order sliding mode is proposed in Sect. 2.2 after introduction in Sect. 2.1. The discrete-time higher order sliding mode control is proposed in Sect. 2.3. The simulation and experimental results are presented in Sect. 2.4. The comparison of the proposed

© Springer Nature Switzerland AG 2019

N. K. Sharma and J. Sivaramakrishnan, *Discrete-Time Higher Order Sliding Mode*,
https://doi.org/10.1007/978-3-030-00172-8_2

sliding mode algorithm with existing algorithms is presented in Sect. 2.5 following with the conclusion in Sect. 2.6.

2.2 Discrete-Time Higher Order Sliding Mode

Consider a generalized discrete-time dynamical system

$$x(k+1) = F[x(k)], x \in \mathbb{R}^n, \tag{2.1}$$

and a sliding function is defined as

$$s(k) = G[x(k)], s \in \mathbb{R}, \tag{2.2}$$

such that it has stable dynamics in the vicinity of the sliding surface $s(k) = 0$.

The finite-difference method is the standard notion to deal with the discrete-time signals [11]. The i-th order forward difference of variable $s(k)$ is defined as

$$\Delta^i s(k) = \sum_{n=0}^{i} (-1)^n \binom{i}{n} s(k+i-n).$$

A formal definition of discrete-time higher order sliding mode in system (2.1) is proposed utilizing the concept of the difference equations.

Definition 2.1 *Discrete-time higher order sliding mode*: A discrete-time higher order sliding mode is defined to take place in a set

$$\mathcal{M} = \left\{ x : \sqrt{\sum_{i=0}^{r-1} \left(\Delta^i s(x) \right)^2} \leq B_d \right\} \tag{2.3}$$

containing the manifold

$$\sigma = \{ x : s = \Delta s = \Delta^2 s = \dots = \Delta^{r-1} s = 0 \},$$

if there exists an open neighbourhood $\mathcal{U}$ of the set $\mathcal{M}$ such that for $x \in \mathcal{U}$, it follows $F[x] \in \mathcal{M}$ and once the trajectory enters the set $\mathcal{M}$, it will always remain in $\mathcal{M}$. Here, r is known as sliding order [12] and B_d is a constant. The set

$$s = \Delta s = \Delta^2 s = \dots = \Delta^{r-1} s = 0, \tag{2.4}$$

is called discrete sliding set.

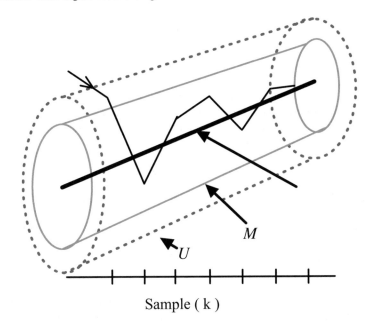

Sample (k)

Fig. 2.1 Pictorial representation of DHOSM

A pictorial representation of the above definition is presented in the Fig. 2.1. The manifold σ is contained in the set $\mathcal{M}$, which is surrounded by the open neighbourhood $\mathcal{U}$. The definition states that once system's state trajectory enters $\mathcal{U}$, it enters into the $\mathbb{R}^r$ dimensional set $\mathcal{M} \subset \mathcal{U}$ in the next sample and then always remains in it. The motion in the set $\mathcal{M}$ is called discrete-time higher order sliding mode.

If trajectory moves perfectly on the manifold σ, then the motion is called ideal discrete-time higher order sliding mode [13]. However, this situation is nearly impossible in discrete-time systems. Therefore, due to the finite switching frequency constraint, discrete-time higher order sliding mode is the practical discrete-time higher order sliding mode. Definition (2.1) allows us to design a discrete-time higher order sliding mode (DHOSM) control.

2.3 Discrete-Time Higher Order Slding Mode Control Design for an LTI System

In this section, a generalized reaching law for r-order sliding mode is proposed. Then, using the proposed reaching law, a discrete-time second-order sliding mode (DSOSM) control is designed for a discrete-time system in the absence of the disturbance. Subsequently, the DHOSM control design is extended for an uncertain discrete-time system considering the presence of the disturbance.

Consider a discrete-time representation of an uncertain LTI system

$$x(k+1) = Ax(k) + Bu(k) + d_x(k), \tag{2.5}$$

where $x \in \mathbb{R}^n$ is the state vector, $u \in \mathbb{R}$ is the control input, and the matrices A and B are known and of appropriate dimensions. The disturbance vector $d_x(k)$ represents the external disturbance effect in the system. The following assumptions are made regarding the disturbance in the system.

Assumption 1 The disturbance vector $d_x(k)$ is matched.

Assumption 2 $d_x(k)$ is norm bounded and its bound is known.

Definition 2.2 *Matched Uncertainty* [14]: An uncertain system (2.5) is said to have a matched uncertainty if the condition $d_x(k) \in$ Range(B) is satisfied. In physical terms, this assumption means that the disturbance affecting the system is entering through the input channel only.

Remark 2.1 In the literature of sliding mode, the effect of unmodeled dynamics and external disturbance are generally assumed to be bounded by a constant [15]. This assumption is not very conservative as a large class of systems fall into this category [16].

Since the sliding surface $s = 0$ governs the closed-loop dynamics of the system, the structure of the sliding function s is chosen appropriately. There are various methods of designing a sliding function [17]. However, typically, the sliding function for an LTI system is chosen to be a linear function. Let us define a sliding function

$$s(k) = cx(k), \tag{2.6}$$

where $s \in \mathbb{R}$ and $cB \neq 0$. The aim is to design a control input for the uncertain system (2.5) such that all the trajectories of (2.5) converge in the set $\mathcal{M}$, as defined in Definition 2.1, and thereafter, always remain in it.

The reaching law approach [5] can be used for designing a DHOSM control. As explained in Chap. 1, the reaching law directly dictates the dynamics of the sliding function. Therefore, a reaching law is designed in such a way that the definition of sliding mode is always satisfied and then, the corresponding control input is synthesized from the reaching law. The reaching law presented by Gao et al. [5] is of the first order, however in order to achieve r-order sliding mode, a generalized reaching law of the order r is required.

A generalized reaching law of order r can be formulated in the form

$$s(k+r) = f(s(k), s(k+1), \ldots, s(k+r-1)), \tag{2.7}$$

such that the sliding function has the desired dynamics.

Similar to the conventional continuous-time higher order sliding mode, the discrete-time higher order sliding mode control is designed according to the relative degree of the sliding function (2.6) with respect to control input u [18]. A sliding function is known to have a relative degree r with respect to the control input u if the control input appears first time only in the r-order difference of s [19, p. 139]. Mathematically, the coefficients of vector $c \in \mathbb{R}^{1 \times n}$ are such that $cB = cAB = cA^2 B = \ldots = cA^{r-2} B = 0$ and $cA^{r-1} B \neq 0$.

Analogous to continuous-time second-order sliding mode control [18], discrete-time second-order sliding mode control can also be designed for two cases:

Case A: *Relative Degree One*

In this case, $cB \neq 0$, i.e. there is a direct relation between the sliding function $s(k+1)$ and the control input $u(k)$. For this case, a generalized reaching law of order two and relative degree one can be designed through the backward expression of (2.7) as

$$s(k+1) = f(s(k-1), s(k)). \tag{2.8}$$

Case B: *Relative Degree Two*

In this case, $cB = 0$ and $cAB \neq 0$, i.e. the control input $u(k)$ has a direct relation with the sliding function $s(k+2)$. For this case, the generalized reaching law has the form of

$$s(k+2) = f(s(k), s(k+1)). \tag{2.9}$$

In this chapter, let us consider that the sliding function (2.6) has the relative degree one. The sliding function with relative degree r, where $r \geq 2$, will be considered in Chap. 4.

2.3.1 Design of Discrete-Time Second-Order Sliding Mode Control in the Absence of Disturbance

The system (2.5) in the absence of disturbance is obtained as

$$x(k+1) = Ax(k) + Bu(k). \tag{2.10}$$

To design DHOSM control for the system (2.10), consider the reaching law (2.8) in the form of

$$s(k+1) = k_1 s(k-1) + k_2 s(k) - \varepsilon_1 T^2 \mathrm{sgn}[s(k-1)] - \varepsilon_2 T^2 \mathrm{sgn}[s(k)], \tag{2.11}$$

where k_1, k_2, ε_1, and ε_2 are the design parameters. The properties of the sliding function dictated by the proposed second-order reaching law (2.11) are discussed in the following theorem.

Definition 2.3 *Ultimately Bounded* [20]: A discrete-time system is said to be ultimately bounded in the set $\mathcal{M}$ iff for every initial condition $x(0)$, we have a finite k^*, such that $x(k) \in \mathcal{M} \; \forall \, k \geq k^*$.

Theorem 2.1 *The sliding function satisfying the reaching law (2.11) with the parameters $0 < k_1 < k_2 < 1$ and $k_1 + k_2 < 1$ is ultimately bounded in a band. Subsequently, the discrete-time higher order sliding mode takes place in a set $\mathcal{M}$, as defined in Definition 2.1.*

Proof The ultimate boundedness of the sliding function can be proved by the Lyapunov-like stability analysis. Let $z_1(k) = s(k-1)$ and $z_2(k) = s(k)$. Now, consider a Lyapunov candidate function

$$V(k) = z_1^2(k) + \alpha z_2^2(k),$$

where α is a positive constant. The first difference of $V(k)$ is computed as

$$\triangle V(k) = V(k+1) - V(k) \tag{2.12}$$
$$= z_1^2(k+1) + \alpha z_2^2(k+1) - z_1^2(k) - \alpha z_2^2(k) \tag{2.13}$$

Consider

$$\delta = \sup \left(\varepsilon_1 T^2 \text{sgn}[s(k-1)] + \varepsilon_2 T^2 \text{sgn}[s(k)] \right) = \varepsilon_1 T^2 + \varepsilon_2 T^2, \tag{2.14}$$

be the upper bound of the switching part of the reaching law (2.11). On using (2.11) and (2.14) in (2.13), we get

$$\triangle V(k) = -[1 - \alpha k_1^2] z_1^2(k) - [\alpha - 1 - \alpha k_2^2] z_2^2(k) + \alpha \delta^2 + 2\alpha \delta k_1 z_1(k) + 2\alpha \delta k_2 z_2(k) +$$
$$2k_1 k_2 z_1(k) z_2(k). \tag{2.15}$$

In (2.15), let $K_a = [1 - \alpha k_1^2]$ and $K_b = [\alpha - 1 - \alpha k_2^2]$, where the parameter α is chosen such that $K_a, K_b > 0$, i.e. α is within the range $\frac{1}{1-k_2^2} < \alpha < \frac{1}{k_1^2}$.

$$\triangle V(k) = -K_a z_1^2(k) - K_b z_2^2(k) + \alpha \delta^2 + 2\alpha \delta k_1 z_1(k) + 2\alpha \delta k_2 z_2(k) + 2k_1 k_2 z_1(k) z_2(k). \tag{2.16}$$

On using Young's inequality [21], $2z_1(k)z_2(k) \leq [z_1^2(k) + z_2^2(k)]$, (2.16) is deduced to be

$$\triangle V(k) \leq -K_a z_1^2(k) - K_b z_2^2(k) + \alpha k_1 k_2 [z_1^2(k) + z_2^2(k)] + \alpha \delta^2 + 2\alpha \delta k_1 z_1(k) + 2\alpha \delta k_2 z_2(k). \tag{2.17}$$

K_a and K_b can be separated such that $K_a = K_{a_1} + K_{a_2}$, $K_b = K_{b_1} + K_{b_2}$, $K_{a_1} > \alpha k_1 k_2$, and $K_{b_1} > \alpha k_1 k_2$. With these parameters,(2.17) can be expanded as

$$\triangle V(k) \le -K_{a_1} z_1^2(k) - K_{b_1} z_2^2(k) + \alpha k_1 k_2 [z_1^2(k) + z_2^2(k)] - \left[\sqrt{K_{a_2}} z_1(k) - \frac{\alpha \delta k_1}{\sqrt{K_{a_2}}} \right]^2 -$$

$$\left[\sqrt{K_{b_2}} z_2(k) - \frac{\alpha \delta k_2}{\sqrt{K_{b_2}}} \right]^2 + \alpha \delta^2 + \frac{\alpha^2 \delta^2 k_1^2}{K_{a_2}} + \frac{\alpha^2 \delta^2 k_2^2}{K_{b_2}}$$

$$\le -K_{a_1} z_1^2(k) - K_{b_1} z_2^2(k) + \alpha k_1 k_2 [z_1^2(k) + z_2^2(k)] + \alpha \delta^2 + \frac{\alpha^2 \delta^2 k_1^2}{K_{a_2}} + \frac{\alpha^2 \delta^2 k_2^2}{K_{b_2}}$$

$$(2.18)$$

On considering

$$g = \alpha \delta^2 + \frac{\alpha^2 \delta^2 k_1^2}{K_{a_2}} + \frac{\alpha^2 \delta^2 k_2^2}{K_{b_2}},$$

$$\beta = \min \left[\left[K_{a_1} - \alpha k_1 k_2 \right], \left[\frac{K_{b_1} - \alpha k_1 k_2}{\alpha} \right] \right],$$

Eq. (2.18) is obtained to be

$$\triangle V(k) \le - \left[K_{a_1} - \alpha k_1 k_2 \right] z_1^2(k) - \left[\frac{K_{b_1} - \alpha k_1 k_2}{\alpha} \right] \alpha z_2^2(k) + g.$$

If the positive design parameter α is chosen such that $g > 0$ and $0 < \beta < 1$, then

$$\triangle V(k) \le -\beta V(k) + g.$$

In this manner, s is found to be ultimately bounded. Subsequently, the ultimate band of s can be found out to be

$$|s| \le \sqrt{\frac{g/\beta}{\alpha}}. \tag{2.19}$$

On using g in (2.19),

$$|s| \le \delta \sqrt{\frac{\left(1 + \frac{\alpha k_1^2}{K_{a_2}} + \frac{\alpha k_2^2}{K_{b_2}} \right)}{\beta}}. \tag{2.20}$$

Thus, the sliding function in the reaching law (2.11) is ultimately bounded in a band

$$|s| \le \beta_d (\varepsilon_1 + \varepsilon_2) T^2,$$

where

$$\beta_d = \sqrt{\frac{\left(1 + \frac{\alpha k_1^2}{K_{a_2}} + \frac{\alpha k_2^2}{K_{b_2}} \right)}{\beta}}.$$

The above Lyapunov stability analysis proves that $s(k)$ and $s(k-1)$ are ultimately bounded in band, therefore,

$$\sqrt{(s(k))^2 + (\Delta s(k))^2} \le (\varepsilon_1 + \varepsilon_2)T^2\sqrt{5}\beta_d,$$

will also be bounded. As the system dynamics are stable confined to the sliding manifold, state trajectory is also ultimately bounded in a set $\mathcal{M}$ and satisfy Definition 2.1. Therefore, the DHOSM takes place in the set $\mathcal{M}$. This completes the proof.

On using reaching law (2.11) for the sliding function (2.6), the control input for the system (2.10) can be synthesized to be

$$u(k) = -(cB)^{-1}\left(cAx(k) - k_1 s(k-1) - k_2 s(k) + \varepsilon_1 T^2 \text{sgn}[s(k-1)] + \varepsilon_2 T^2 \text{sgn}[s(k)]\right).$$
(2.21)

2.3.2 Design of Discrete-Time Second-Order Sliding Mode Control in the Presence of Disturbance

For the uncertain LTI system (2.5), let us define the effect of the disturbance vector on the sliding function as

$$d(k) = cd_x(k). \tag{2.22}$$

Since d_x is assumed to be bounded, there exists a known constant d^* such that $|d(k)| \le d^*$. Further, there exist known d_u and d_l such that

$$d_l \le d(k) \le d_u, \tag{2.23}$$

where d_u and d_l denote the upper and lower bounds of $d(k)$ respectively. The mean of disturbance bounds is defined as

$$d_0 = \frac{d_u + d_l}{2}.$$

The reaching law in the presence of the disturbance is proposed as

$$s(k+1) = k_1 s(k-1) + k_2 s(k) - \varepsilon_2 T^2 \text{sgn}[s(k)] - \varepsilon_1 T^2 \text{sgn}[s(k-1)] + d(k) - d_0. \tag{2.24}$$

Theorem 2.2 proves the ultimate boundedness of the sliding function dictated by the reaching law (2.24).

Theorem 2.2 *The sliding function satisfying the reaching law (2.24) with the parameters* $0 < k_1 < k_2 < 1$, $k_1 + k_2 < 1$, *and* $(\varepsilon_1 T^2 + \varepsilon_2 T^2) \geq |d^* - d_0|$ *is ultimately bounded in a band. Subsequently, the discrete-time higher order sliding mode takes place in a set $\mathcal{M}$, as defined in Definition 2.1.*

Proof The supremum of the switching part in the presence of disturbance is

$$\delta_d = \sup \left(\varepsilon_1 T^2 \text{sgn}[s(k-1)] + \varepsilon_2 T^2 \text{sgn}[s(k)] + d(k) - d_0 \right)$$
$$= \varepsilon_1 T^2 + \varepsilon_2 T^2 + |d^* - d_0|. \tag{2.25}$$

The proof of ultimate boundedness of the sliding function, dictated by the reaching law (2.24), in the DHOSM band

$$|s(k)| \leq \beta_d((\varepsilon_1 + \varepsilon_2)T^2 + |d^* - d_0|), \tag{2.26}$$

is similar to the proof of Theorem 2.1. The only difference is that the bound δ_d is used in the place of δ. To ensure the ultimate boundedness of the sliding function in DHOSM band, the switching part of reaching law should be greater or equal to the maximum of the disturbance term [22], i.e.

$$(\varepsilon_1 T^2 + \varepsilon_2 T^2) \geq |d^* - d_0|. \tag{2.27}$$

Similar to Theorem 2.1, the set $\sqrt{(s(k))^2 + (\Delta s(k))^2}$ will also be bounded in a band and as the system dynamics are stable confined to the sliding manifold, state trajectory will also be ultimately bounded in a set $\mathcal{M}$. Thus, Definition 2.1 is satisfied and DHOSM takes place in the set $\mathcal{M}$. This completes the proof.

For the system (2.5), the control input in the presence of disturbance can be synthesized to be

$$u(k) = -(cB)^{-1} \left(cAx(k) - k_1 s(k-1) - k_2 s(k) + \varepsilon_1 T^2 \text{sgn}[s(k-1)] + \varepsilon_2 T^2 \text{sgn}[s(k)] \right) - (cB)^{-1} d_0. \tag{2.28}$$

The control algorithm proposed in the paper is numerically and experimentally validated on a rectilinear plant and the results are presented in the following section.

2.4 Numerical and Experimental Validation

2.4.1 Description of the Experimental Setup

The rectilinear plant, as shown in Fig. 2.2, is an electromechanical system that represents physical plants with rigid bodies, flexible linear drives with gear and belt arrangement, and discrete vibration coupling with actuator at the drive input and sensor. This plant is very susceptible to external disturbance and parameter variation and thus, it is an appropriate setup to verify the robustness of the proposed and existing algorithms. This setup is manufactured by Educational Control Products (ECP), USA [23]. The plant consists of three serially connected mass–spring–damper systems. The control input is applied through brushless DC servomotor. Three high-resolution encoders are utilized to measure the position of the masses. A separate servomotor is also attached to provide external disturbance to the system.

In this chapter, a single mass–spring system is chosen for the experiment. The free body diagram of the single mass–spring system is shown in Fig. 2.3. The dynamics of the single mass–spring system can be represented by a second-order discrete-time equation (2.5) with parameters $A = e^{A_c T}$ and $B = \left(\int_0^T e^{A_c \tau} d\tau \right) B_c$, where $A_c = \begin{bmatrix} 0 & 1 \\ -\frac{k}{m} & -\frac{c}{m} \end{bmatrix}$ and $B_c = \begin{bmatrix} 0 \\ \frac{1}{m} \end{bmatrix}$, and the values of parameters are given in Table 2.1.

Fig. 2.2 Experiment setup of the rectilinear plant

Fig. 2.3 Free body diagram
of mass–spring system

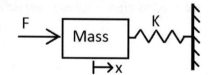

Table 2.1 Parameters value in single mass damper experiment

Parameter	Value
k	737.41 N/m
m	2.77 Kg
c	0 N/(m/s)
T	0.1 s
k_1	0.03
k_2	0.4
ε_1	25
ε_2	25

2.4.2 Simulation Results

The proposed control algorithm is simulated using MATLAB with parameters presented in Table 2.1. The initial condition is chosen to be $x_0 = \begin{bmatrix} 10 & 10 \end{bmatrix}^T$. The simulation results are depicted in Fig. 2.4.

Figure 2.4a shows the evolution of the sliding function. It can be seen that the sliding function converges gradually. The control input, depicted in Fig. 2.4b is also bounded. It should be noted in Fig. 2.4a that the sliding function in steady-state alternates between two extreme value with an intermediate value. The feature is due to the fact that $s(k+1)$ depends on sgn$[s(k)]$ as well as sgn$[s(k-1)]$. On the other hand, in the conventional first-order reaching law based sliding mode presented by Gao et al. in [5], the steady-state value of the $s(k+1)$ depends on sgn$[s(k)]$ alone and alternates between two values in the sliding mode band, which results in chattering with large magnitude. Thus, due to the choice of controller parameters, the proposed control law gives smoother sliding motion and the chattering amplitude is reduced by a considerable amount.

2.4.3 Experimental Results

2.4.3.1 In the Absence of Disturbance

The experimental results are presented in Figs. 2.5a and 2.8. As shown in Fig. 2.5a, the sliding function decreases in the reaching phase and becomes stable in steady state. Then, due to the switching terms, sliding function alternates with intermediate states as shown in the inset. The control input, as shown in Fig. 2.5b, is also bounded in steady state.

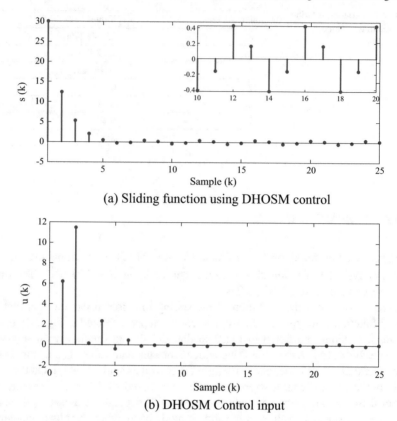

(a) Sliding function using DHOSM control

(b) DHOSM Control input

Fig. 2.4 Simulation results using the proposed DHOSM control algorithm

2.4.3.2   In the Presence of Disturbance

Further, to examine the performance of the proposed control law in the presence of external disturbance, a disturbance motor is employed to provide the external disturbance

$$d_x(k) = \begin{bmatrix} 0 & 0.5 \end{bmatrix}^T \times \sin(k), \tag{2.29}$$

to the system. As shown in Fig. 2.6, in the presence of the external disturbance, the proposed control law attenuates the effect of disturbance.

The control algorithms presented in Bartoszewicz et al. [6] and Fridman et al. [8] are also implemented to compare the performance of the proposed algorithm. The comparative results are shown in Figs. 2.7 and 2.8 and are explained in Sect. 2.5.

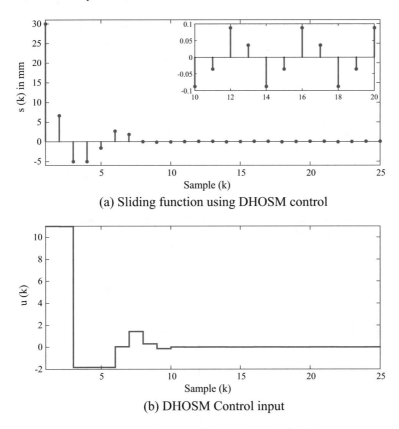

(a) Sliding function using DHOSM control

(b) DHOSM Control input

Fig. 2.5 Experimental results using the proposed DHOSM control algorithm

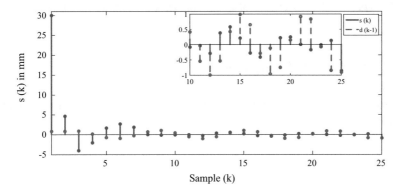

Fig. 2.6 Experimental response of the sliding function in the presence of external disturbance on using the proposed DHOSM control technique

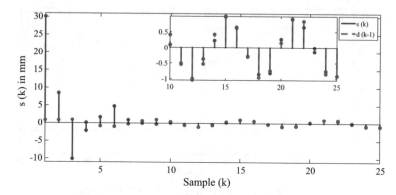

Fig. 2.7 Experimental response of the sliding function in presence of the external disturbance on using control algorithm presented in [6]

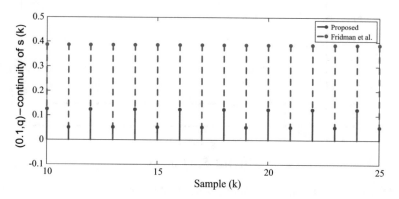

Fig. 2.8 (p, q_r)-continuity of the sliding function when $T = 0.1$ sec. on using the proposed DHOSM control technique

2.5 Performance Comparison

The proposed control algorithm is compared with the existing control algorithms on the basis of sensitivity analysis and (p, q)-continuity.

2.5.1 Sensitivity Analysis

The property of disturbance rejection can be appropriately characterized by the sensitivity function. The sensitivity function of a function $F(k)$ with respect to its parameter η can be defined as

$$\mathbf{S}_\eta^{F(k)} = \frac{\mathrm{d}F(k)}{\mathrm{d}\eta} \times \frac{\eta}{F(k)}. \tag{2.30}$$

Bartoszewicz [6] proposed a control algorithm such that the sliding function in the steady state satisfies

$$|s(k+1)| = |d(k) - d_0|,$$

and its sensitivity function with respect to disturbance $d(k)$ can be obtained as

$$|\mathbf{S}_{d(k)}^{s(k+1)}| = \left| \frac{ds(k+1)}{dd(k)} \times \frac{d(k)}{s(k+1)} \right| = \left| \frac{d(k)}{d(k) - d_0} \right|. \qquad (2.31)$$

Therefore in the band, if d_0 is zero, sensitivity function of sliding mode is equal to 1. This means that the sliding mode control does not attenuate the disturbances and the sliding function is proportional to the disturbance. On the other hand, the sensitivity function of the proposed algorithm (2.24) in the steady state is obtained to be

$$|\mathbf{S}_{d(k)}^{s(k+1)}| = \left| \frac{d(k)}{(\varepsilon_1 T^2 + \varepsilon_2 T^2) + (d^* - d_0)} \right|. \qquad (2.32)$$

As ε_1 and ε_2 are chosen such that the inequality (2.27) is always satisfied, the sensitivity function of the proposed algorithm is always lesser than that of algorithm presented in [6], for any value of d_0, relatively. Moreover, the dependency of the design parameters ε_1 and ε_2 on the sensitivity function provides the potential to tune sensitivity function according to the disturbance.

For the setup and design parameters considered in the experiment, the maximum sensitivity of the conventional first-order sliding mode control [6] in the steady state is 1. While the maximum sensitivity by the proposed second-order sliding mode control is found out to be 0.5, which is a 50% improvement in the disturbance rejection capability relatively. The performed experiment also validates the robustness of the proposed control law. On implementing the control law presented in [6], the sliding function $s(k)$ is found almost equal to the disturbance $d(k-1)$ in the quasi-sliding mode band, as shown in Fig. 2.7. On the other hand, as shown in Fig. 2.6, the proposed control law rejects the disturbance as magnitude of $s(k)$ becomes lesser than the disturbance $d(k-1)$. Hence, the proposed control algorithm has better disturbance rejection capability.

2.5.2 (p, q_r)-continuity

It can be noticed that the definitions of continuous-time sliding mode are based upon the conventional definitions of the continuity. However, these definitions do not hold for discrete-time. Thus, it is necessary to provide definitions in the context of the discrete time. Discrete-time sliding mode can be understood as a discrete set, where sliding function has a certain value at a particular sampling instant. This premise allows to define a relaxed version of continuity in discrete-time higher order sliding function.

Definition 2.4 *Continuity in discrete-time sliding function*: A discrete-time r-order sliding function $s \in \mathbb{R}$ is defined to be (p, q_r)-continuous at a sampling instant $a \in \mathbb{R}$ if $s(a)$ is defined and for any $\varepsilon > 0$ there exists a $\delta > 0$ such that the inequality $|a - k\tau| < p + \delta$ implies the inequality $|s(a) - s(x(k\tau))| < q_r + \varepsilon$, for all $k > 0$, where $p, q_r \in \mathbb{R}$.

Corollary 2.1 *The discrete-time sliding function, $s(k\tau)$, is defined at the sampling instants only. In this situation, p is lower bounded by the sampling instant τ, i.e. $p \geq \tau$. If $p = \tau$, then $q_r = |s((k+1)\tau) - s(k\tau)|$.*

The above definition and corollary clear that lesser the value of (p, q_r)-continuity, closer is the function to conventional continuity. The sliding order of continuous-time higher order sliding mode affects sliding accuracy of real sliding [12]. Similarly, the sliding order affects the (p, q_r)-continuity of the discrete-time sliding function. The relation between the sliding order and (p, q_r)-continuity is presented as the order of the continuity.

Definition 2.5 *Order of the continuity in discrete-time sliding function*: For a r-order discrete-time sliding function to be (p, q_r)-continuous, there must be a q_r such that

$$q_r \leq C|p|^r, \tag{2.33}$$

where C is a positive constant. If $p = \tau$, then q_r must satisfy the inequality

$$q_r \leq C|\tau|^r. \tag{2.34}$$

In this condition, r is called the order of the continuity in discrete-time sliding function.

Proposition 2.1 *If the r-order discrete-time sliding function is (p, q_r)-continuous, then $\triangledown S(k\tau, x(k\tau))$, $\triangledown^2 S(k\tau, x(k\tau))$, ..., $\triangledown^{r-1} S(k\tau, x(k\tau))$ are $(p, q_{r,1})$, $(p, q_{r,2})$, ..., $(p, q_{r,r-1})$-continuous respectively, where $q_{r,1}, q_{r,2}, ..., q_{r,r-1}$ holds following inequalities:*

$$q_{r,1} \leq C_1 \tau^{r-1},$$
$$q_{r,2} \leq C_2 \tau^{r-2},$$
$$q_{r,r-1} \leq C_{r-1} \tau^1,$$

where $C_1, C_2, \ldots, C_{r-1}$ are the positive constants and τ is the sampling time.

Thus, the order of continuity represents the degree of smoothness in discrete-time sliding function. Higher the order of continuity, lesser the value of q, better the smoothness of discrete-time sliding mode.

On comparing the proposed control algorithm with the other existing algorithms on the basis of the (p, q_2)-continuity, the proposed second-order sliding mode control algorithm provides continuity of the order 2. Thus, the proposed control algorithm provides better continuity than the existing first-order sliding mode algorithms [7].

Moreover, the proposed algorithm is compared with a super-twisting like algorithm [8], which is a discretized version of a continuous-time second-order sliding mode. Figure 2.8 presents the comparison between the (p, q_r)-continuity of the proposed sliding mode and sliding mode presented in [8] with the sampling time 0.1 sec. As shown in the figure, super-twisting like algorithm [8] presents a constant value of q_r in steady state. On the other hand, proposed reaching law exhibits lower value of q_r. Thus, the proposed algorithm for second-order sliding mode has better continuity compared to [8]. Therefore, we can conclude that the proposed sliding algorithm improves the (p, q_r)-continuity of the discrete-time sliding mode. This implies a smoother response as compared to a discretized version of continuous-time sliding mode algorithm [8].

2.6 Conclusion

In this chapter, a formal definition of discrete-time higher order sliding mode is proposed. A discrete-time higher order sliding mode control is designed for an uncertain LTI system. The analytic proofs and experimental results validate the definition of discrete-time higher order sliding mode proposed in the chapter. On comparing the proposed sliding mode algorithm with the other existing sliding mode algorithms on the basis of sensitivity function and (p, q_r)-continuity, the results and comparisons show that the proposed sliding mode control has superior disturbance rejection capabilities as well as better continuity.

References

1. J. Zhang, G. Feng, Y. Xia, IEEE Trans. Ind. Electron. **61**(5), 2432 (2014)
2. J. Zhang, Y. Lin, G. Feng, IET Control Theor. Appl. **9**(8), 1205 (2015)
3. N.D. That, Q.P. Ha, IET Control Theor. Appl. **9**, 1700 (2015)
4. J. Hu, Z. Wang, H. Gao, L.K. Stergioulas, IEEE Trans. Ind. Electron. **59**(7), 3008 (2012)
5. W. Gao, Y. Wang, A. Homaifa, IEEE Trans. Ind. Electron. **42**(2), 117 (1995)
6. A. Bartoszewicz, IEEE Trans. Ind. Electron. **45**(4), 633 (1998)
7. S. Chakrabarty, B. Bandyopadhyay, Automatica **52**, 83 (2015)
8. L. Fridman, O. Camacho, I. Chairez, B. Bandyopadhyay, I. Salgado, IET Control Theor. Appl. **10**(8), 803 (2014)
9. B. Wang, X. Yu, X. Li, IEEE Trans. Ind. Electron. **55**(11), 4055 (2008)
10. B. Wang, X. Yu, G. Chen, Automatica **45**(1), 118 (2009)

11. G.F. Franklin, D. Powell, M. Workman, *Digital Control of Dynamic Systems*, 3nd edn. (Addison Wesley, 1997)
12. A. Levant, Int. J. Control **58**(6), 1247 (1993)
13. N.K. Sharma, S. Singh, S. Janardhanan, *in International Workshop on Recent Advances in Sliding Modes (RASM)* (Turkey, Istanbul, 2015), pp. 1–6
14. S. Janardhanan, V. Kariwala, IEEE Trans. Autom. Control **53**(1), 367 (2008)
15. A. Argha, L. Li, S.W. Su, H. Nguyen, in *53rd IEEE Conference on Decision and Control* (2014), pp. 4747–4752
16. Y. Xia, G.P. Liu, P. Shi, J. Chen, D. Rees, J. Liang, IET Control Theor. Appl. **1**(4), 1169 (2007)
17. C. Edwards, S. Spurgeon, *Sliding Mode Control: Theory And Applications* (CRC Press, 1998)
18. A. Pisano, Second order sliding modes: Theory and applications. Ph.D. thesis, University of Cagliari, Italy (2000)
19. A. Isidori, *Nonlinear Control Syst.*, 3rd edn. (Springer, London, 1995)
20. F. Blanchini, IEEE Trans. Autom. Control **39**(2), 428 (1994)
21. A. Jain, S. Bhasin, in *IEEE Conference on Control Applications* (Sydney, Australia, 2015), pp. 1686–1691
22. S. Qu, X. Xia, J. Zhang, IEEE Trans. Ind. Electr. **61**(7), 3502 (2014)
23. *Installation and User Manual of ECP Model 210 for Use With MATLAB-R14 using Real Time Windows Target (RTWT), Version 4.0* (Educational Control Products, Woodland Hills, CA, USA, 2004)

Chapter 3
Optimal Discrete-Time Higher Order Sliding Mode

3.1 Introduction

The choice of sliding surface plays an important role in the performance of sliding mode as it defines the dynamics of the closed-loop system. Once the system's state trajectories are on the sliding surface, the behaviour of the system governs by the sliding surface. Thus, the design of the sliding surface is a crucial step in the sliding mode control design [1]. There exist multiple methods to design a sliding surface [2]. Among these methods, an optimal sliding surface design method has been a major focus of research [3–5]. In optimal sliding mode, the sliding surface is designed such that a predefined performance index is minimized such that system's state as well as control input is optimized. To design such sliding surfaces, various performance indices are used such as linear quadratic function (LQ) [6], Bolza–Meyer criterion based non-quadratic function [7], and integral time-weighted absolute error (ITAE) based function [8]. The design strategies for optimal sliding surface have been well studied for continuous-time higher order sliding mode [9–11] and discrete-time first-order sliding mode [1, 12–14].

However, an optimal higher order sliding surface has not yet been attempted in discrete-time, to the best of our knowledge. The fact that a linear quadratic function based optimal control optimizes the system within a fixed time motivated us to design an optimal discrete-time higher order sliding mode control such that it guarantees the establishment of DHOSM. Furthermore, in practical stituations, the information of all states is generally not always available or requires many sensors [15]. Therefore, the control input should be designed such that it uses minimal state information.

The major contributions in this chapter are as follows.

1. A time-varying discrete-time higher order sliding surface is designed for an LTI system such that when confined to it, a specified linear quadratic cost function is minimized over a certain time period for fixed final state.
2. An optimal DHOSM control is proposed which ensures that the sliding function and its higher order differences enter into a DHOSM band and remain therein despite the disturbance in the system.

© Springer Nature Switzerland AG 2019

N. K. Sharma and J. Sivaramakrishnan, *Discrete-Time Higher Order Sliding Mode*, https://doi.org/10.1007/978-3-030-00172-8_3

3. The proposed control input utilizes information of only measurable states. Therefore, it is easier and economical to implement experimentally as compared to the
   classical DHOSM.
4. A disturbance estimation technique is used to reduce the DHOSM band further.
5. The proposed control technique is experimentally validated on a rectilinear plant
   setup.

The organization of the chapter is as follows. Section 3.2 introduces the Optimal
DHOSM control problem. In Sect. 3.3, a control input is designed first, then it is
further modified using disturbance estimation technique. The stability of the system
under consideration using the proposed control is analysed in Sect. 3.4. The simulation and experimental results are presented in Sect. 3.5 followed by conclusions in
Sect. 3.6.

3.2 Problem Formulation

Consider an uncertain discrete-time LTI system in the controllable canonical form
as presented in Chap. 2

$$x(k + 1) = Ax(k) + bu(k) + d_x(k), \tag{3.1}$$

where $x \in \mathbb{R}^n$ is the state vector, $u \in \mathbb{R}$ is the control input, $d_x(k)$ is the disturbance
vector, and $A = \begin{bmatrix} 0 & 1 & 0 & \cdots & 0 \\ 0 & 0 & 1 & \cdots & 0 \\ \vdots & \vdots & \vdots & \ddots & \vdots \\ -a_1 & -a_2 & -a_3 & \cdots & -a_n \end{bmatrix}$ and $b = \begin{bmatrix} 0 \\ 0 \\ \vdots \\ 1 \end{bmatrix}$.

Assumption 3 The disturbance vector $d_x(k)$ is matched, i.e. $d_x(k) \in$ Range (b). The
disturbance vector $d_x(k)$ is assumed to be norm bounded by a known constant.

Assumption 4 The states x_i for $i = (1, \ldots, n - r + 1)$ are measurable and remain-
ing $r - 1$ states are observable but not measurable.

A sliding function is defined as

$$s(k) = cx(k) = \sum_{i=1}^{n-r} c_i x_i(k) + x_{n-r+1}, \tag{3.2}$$

where $s \in \mathbb{R}$ and vector $c = [c_1, c_2, c_3, \ldots, c_{n-r-1}, c_{n-r}, 1, 0, \ldots, 0] \in \mathbb{R}^n$ such that
system (3.1) has stable dynamics when confined to sliding surface. It is assumed that
sliding surface s has relative degree r with respect to input u. A higher order sliding
surface is defined as

$$\bar{s}(k) = \bar{c}\xi(k) \tag{3.3}$$

where $\xi(k) = \left[s_1(k) \, s_2(k) \cdots s_r(k) \right]^T \in \mathbb{R}^r$, $s_i(k) = s(k+i-1)$ for $i = (1, \ldots, r)$, and vector $\bar{c} = [\bar{c}_1, \bar{c}_2, \bar{c}_3, \ldots, \bar{c}_{r-1}, 1] \in \mathbb{R}^r$ contains coefficients such that higher order sliding surface $\bar{s}$ has stable dynamics.

The DHOSM is defined in Chap. 2 using the higher order differences of the sliding function. However, a difference equation can be represented using either shift operator or delta operator [16]. Both the forms represent the same difference equation and, for the sake of simplicity, can be transformed into each other using relation

$$E^i s(k) = s(k+i) = \sum_{n=0}^{i} \binom{i}{n} \Delta^n s(k) \tag{3.4}$$

and

$$\Delta^i s(k) = \sum_{n=0}^{i} (-1)^n \binom{i}{n} s(k+i-n). \tag{3.5}$$

It should be noted that the discrete sliding set (2.4) is defined in 2.2 as

$$s(k) = \Delta s(k) = \Delta^2 s(k) = \ldots = \Delta^{r-1} s(k) = 0.$$

Using the relation (3.4) and (3.5), discrete sliding set (2.4) can also be represented as

$$s(k) = s(k+1) = s(k+2) = \ldots = s(k+r-1) = 0. \tag{3.6}$$

Similarly, DHOSM can also be defined using the sliding set (3.6) and shift operators.

Definition 3.1 *Discrete-time higher order sliding mode*: For an uncertain discrete-time system $x(k+1) = F[x(k)]$ and a sliding function $s(k) = H[x(k)]$, a discrete-time higher order sliding mode is defined to take place in a set

$$\mathcal{M}_s = \left\{ x : \sqrt{\sum_{i=0}^{r-1} (s(k+i))^2} \le B_s \right\} \tag{3.7}$$

containing the manifold

$$\sigma = \{x : s(k) = s(k+1) = s(k+2) = \ldots = s(k+r-1) = 0\},$$

if there exists an open neighbourhood $\mathcal{U}_s$ of the set $\mathcal{M}_s$ such that for $x \in \mathcal{U}_s$, it follows $F[x] \in \mathcal{M}_s$ and once the trajectory enters the set $\mathcal{M}_s$, it will always remain in $\mathcal{M}_s$. Here, r is known as sliding order [17] and B_s is a constant.

The states of the system can be partitioned into two parts $\bar{x} = \begin{bmatrix} x_1, x_2, \ldots, x_{n-r} \end{bmatrix}^T$ $\in \mathbb{R}^{n-r}$ and $\underline{x} = \begin{bmatrix} x_{n-r+1}, x_{n-r+2}, \ldots, x_n \end{bmatrix}^T \in \mathbb{R}^r$, and the system model (3.1) can be transformed into a new form with state $z = [\bar{x}, \xi] \in \mathbb{R}^n$ through a state transformation

$$z(k) = Gx(k), \tag{3.8}$$

where

$$G = \begin{bmatrix} I_{n-r} & 0 \\ G_1 & G_2 \end{bmatrix}, \tag{3.9}$$

$$G_1 = \begin{bmatrix} c_1 & c_2 & c_3 & \cdots & c_{n-r} \\ 0 & c_1 & c_2 & \cdots & c_{n-r-1} \\ 0 & 0 & c_1 & \cdots & c_{n-r-2} \\ \vdots & \vdots & \vdots & \ddots & \vdots \\ 0 & 0 & 0 & \cdots & \cdots \end{bmatrix}, \tag{3.10}$$

and

$$G_2 = \begin{bmatrix} 1 & 0 & \cdots & 0 & 0 \\ c_{n-r} & 1 & \cdots & 0 & 0 \\ \vdots & \vdots & \ddots & \vdots & \vdots \\ \cdots & \cdots & c_{n-r} & 1 & 0 \\ \cdots & \cdots & c_{n-r-1} & c_{n-r} & 1 \end{bmatrix}. \tag{3.11}$$

Thus, the transformed system is obtained to be

$$z(k+1) = \bar{A}z(k) + \bar{b}u(k) + d(k), \tag{3.12}$$

where $\bar{A} = GAG^{-1}$, $\bar{b} = Gb = \begin{bmatrix} \bar{b}_1 & \bar{b}_2 \end{bmatrix}^T$ and $d = Gd_x = \begin{bmatrix} d_{\bar{x}} & d_{\xi} \end{bmatrix}^T$. The system (3.12) can be partitioned as

$$\begin{bmatrix} \bar{x}(k+1) \\ \xi(k+1) \end{bmatrix} = \begin{bmatrix} \bar{A}_{11} & \bar{A}_{12} \\ \bar{A}_{21} & \bar{A}_{22} \end{bmatrix} \begin{bmatrix} \bar{x}(k) \\ \xi(k) \end{bmatrix} + \begin{bmatrix} 0 \\ \bar{b}_2 \end{bmatrix} u(k) + \begin{bmatrix} 0 \\ d_{\xi}(k) \end{bmatrix}, \tag{3.13}$$

where

$$\bar{A}_{11} = \begin{bmatrix} 0 & 1 & 0 & \cdots & 0 & 0 \\ 0 & 0 & 1 & \cdots & 0 & 0 \\ \vdots & \vdots & \vdots & \ddots & \vdots & \vdots \\ 0 & 0 & 0 & \cdots & 0 & 1 \\ -c_1 & -c_2 & -c_3 & \cdots & -c_{n-r-1} & -c_{n-r} \end{bmatrix},$$

$$\bar{A}_{12} = \begin{bmatrix} 0 & 0 & \cdots & 0 & 0 \\ 0 & 0 & \cdots & 0 & 0 \\ \vdots & \vdots & \ddots & \vdots & \vdots \\ 0 & 0 & \cdots & 0 & 0 \\ 1 & 0 & \cdots & 0 & 0 \end{bmatrix},$$

$$\bar{A}_{21} = \begin{bmatrix} 0 & 0 & \cdots & 0 \\ 0 & 0 & \cdots & 0 \\ \vdots & \vdots & \ddots & \vdots \\ 0 & 0 & \cdots & 0 \\ -\bar{a}_1 & -\bar{a}_2 & \cdots & -\bar{a}_{n-r} \end{bmatrix},$$

$$\bar{A}_{22} = \begin{bmatrix} 0 & 1 & 0 & \cdots & 0 & 0 \\ 0 & 0 & 1 & \cdots & 0 & 0 \\ \vdots & \vdots & \vdots & \ddots & \vdots & \vdots \\ 0 & 0 & 0 & \cdots & 0 & 1 \\ -\bar{a}_{n-r+1} & -\bar{a}_{n-r+2} & -\bar{a}_{n-r+3} & \cdots & -\bar{a}_{n-1} & -\bar{a}_n \end{bmatrix},$$

$\bar{b}_2 = \begin{bmatrix} 0 & \cdots & 1 \end{bmatrix}^T \in \mathbb{R}^r$, and $d_\xi \in \mathbb{R}^r$.

The objective is to design an optimal higher order sliding surface and its corresponding optimal control law for system (3.12) such that sliding function s and its higher order terms are stabilized in a band and DHOSM, as defined in Definition 3.1, can be achieved.

3.3 Optimal DHOSM Control Design

3.3.1 Optimal DHOSM Control Design in Absence of Disturbance in the System

On considering no disturbance in the system, the control input includes the equivalent control part and switching control part. The equivalent control is designed through optimal control formulation with partial state information.

3.3.1.1 Equivalent Control Design

Consider the subsystem of (3.13) with equivalent control in the absence of disturbance,

$$\xi(k+1) = \bar{A}_{21}x(k) + \bar{A}_{22}\xi(k) + \bar{b}_2 u_{eq}(k). \tag{3.14}$$

Due to partial state information, the equivalent control u_{eq} contains two parts: a direct feedback control u_f and an optimal control u_{opt}, i.e.

$$u_{eq}(k) = u_f(k) + u_{opt}(k).$$

As it is assumed that only $(n - r + 1)$ states are measurable, the measurable states can be used to design the direct feedback control part as

$$u_f(k) = -\left(\bar{c}\bar{b}_2\right)^{-1}\left(\bar{c}\bar{A}_{21}\right)\bar{x}(k) = \sum_{i=1}^{n-r}\bar{a}_i x_i(k). \qquad (3.15)$$

On the other hand, the optimal control part u_{opt} is to be designed such that it minimizes a predefined objective function in finite time. On utilizing the feedback control input (3.15), the subsystem (3.24) reduces to

$$\xi(k + 1) = \bar{A}_{22}\xi(k) + \bar{b}_2 u_{opt}(k). \qquad (3.16)$$

The aim is to design the higher order sliding surface and its corresponding controller, such that when the closed-loop system is confined to the higher order sliding surface, the quadratic performance index

$$J = \sum_{k=0}^{k_f-1}\left[\xi^T(k)Q\xi(k) + u_{opt}^T(k)Ru_{opt}(k)\right] \qquad (3.17)$$

is minimized in the finite time. The system (3.16) can be further partitioned with variables $\xi_1 = [s_1, s_2, \ldots, s_{r-1}]^T \in \mathbb{R}^{r-1}$ and $\xi_2 = s_r \in \mathbb{R}$ as

$$\xi_1(k + 1) = \Phi_{11}(k)\xi_1(k) + \Phi_{12}(k)\xi_2(k),$$
$$\xi_2(k + 1) = \Phi_{21}(k)\xi_1(k) + \Phi_{22}(k)\xi_2(k) + u_{opt}(k),$$

where $\Phi_{11} = \begin{bmatrix} 0 & 1 & 0 & \cdots & 0 \\ 0 & 0 & 1 & \cdots & 0 \\ \vdots & \vdots & \vdots & \ddots & \vdots \\ 0 & 0 & 0 & \cdots & 1 \end{bmatrix} \in \mathbb{R}^{(r-1)\times(r-1)}$, $\Phi_{12} = \begin{bmatrix} 0 \\ 0 \\ \vdots \\ 1 \end{bmatrix} \in \mathbb{R}^{(r-1)\times 1}$,

$\Phi_{21} = [-\bar{a}_{n-r+1}, -\bar{a}_{n-r+2}, \ldots, -\bar{a}_{n-1}] \in \mathbb{R}^{1\times(r-1)}$, and $\Phi_{22} = -\bar{a}_n \in \mathbb{R}$.
The higher order sliding surface is defined as

$$\bar{s}(k) = \begin{bmatrix} K(k) & 1 \end{bmatrix}\begin{bmatrix} \xi_1(k) \\ \xi_2(k) \end{bmatrix}, \qquad (3.18)$$

where $K = [\bar{c}_1, \bar{c}_2, \bar{c}_3, \ldots, \bar{c}_{r-1}] \in \mathbb{R}^{r-1}$. The following theorem is stated to obtain an optimal higher order sliding surface and its corresponding optimal control input.

> **Theorem 3.1** *The optimal time-varying higher order sliding surface that leads to minimization of the quadratic performance index in (3.17) within finite sample k_f is of the form $\bar{s}(k) = \begin{bmatrix} K(k) & 1 \end{bmatrix} \xi(k)$, where $K(k)$ is the solution of*
>
> $$K(k)\Phi_{12}K(k) + (\Phi_{22} + F_2(k))K(k) - K(k)\Phi_{11} - (F_1(k) + \Phi_{21}) = 0,$$
>
> *at sample k with $u_{opt}(k) = \begin{bmatrix} F_1(k) & F_2(k) \end{bmatrix} \xi(k)$ being the LQR optimal state feedback control for minimizing (3.17).*

Proof The optimal control that minimizes the cost function (3.17) can be found to be [18, p. 209]

$$u_{opt}(k) = -R^{-1}\bar{b}_2^T A_{22}^{-T} [W(k) - Q]\xi(k),$$

where W is the solution of the matrix difference Riccati equation (DRE)

$$W(k) = \bar{A}_{22}^T W(k+1) \left[I + \bar{b}_2 R^{-1} \bar{b}_2^T W(k+1)\right]^{-1} \bar{A}_{22} + Q,$$

which can be solved backward in time. The final state is chosen to be zero and it is assumed that the inversion of $W(k)$ exists for all $k \leq k_f$ [18]. The optimal control can be represented as

$$u_{opt}(k) = \begin{bmatrix} F_1(k) & F_2(k) \end{bmatrix} \begin{bmatrix} \xi_1(k) \\ \xi_2(k) \end{bmatrix}, \tag{3.19}$$

where $F_1 = \begin{bmatrix} f_1 & f_2 & \cdots & f_{r-1} \end{bmatrix} \in \mathbb{R}^{r-1}$ and $F_2 = f_r \in \mathbb{R}$. When the states are confined to the higher order sliding surface,

$$u_{opt}(k) = (F_1(k) - F_2(k)K(k))\,\xi_1(k). \tag{3.20}$$

To achieve and maintain ξ on the higher order sliding surface, $\bar{s}(k+1)$ should be equal to zero. The control input required for $\bar{s}(k+1) = 0$ can be obtained as

$$u(k)|_{\bar{s}(k+1)=0} = -(K(k)\Phi_{11} - K(k)\Phi_{12}K(k) + \Phi_{21} - \Phi_{22}K(k))\xi_1(k). \tag{3.21}$$

For the higher order sliding surface to be optimal, control (3.21) should be equal to the optimal control (3.20). Equating (3.20) and (3.21), we get

$$K(k)\Phi_{12}K(k) + (\Phi_{22} + F_2(k))\,K(k) - K(k)\Phi_{11} - (\Phi_{21} + F_1(k)) = 0. \tag{3.22}$$

The optimal higher order sliding function can then be represented as $\bar{s}(k) = \begin{bmatrix} K(k) & 1 \end{bmatrix} \xi(k)$, where $K(k)$ is the solution of (3.22) at the sample k. The procedure of solving (3.22) is presented in [1]. This completes the proof.

Therefore, the equivalent control which stabilizes the system (3.24) within finite sample $k_f > r - 1$ is obtained to be

$$u_{eq}(k) = u_f(k) + u_{opt}(k),\qquad(3.23)$$

where u_f and u_{opt} are defined in (3.15) and (3.19), respectively.

3.3.2 Optimal DHOSM Control Design in the Presence of Disturbance in the System

Considering the presence of disturbance in the system, the subsystem (3.24) is structured in

$$\xi(k+1) = \bar{A}_{21}x(k) + \bar{A}_{22}\xi(k) + \bar{b}_2 u(k) + d_\xi(k).\qquad(3.24)$$

To compensate the disturbance in the system, the switching control part $u_s(k)$ is designed to be

$$u_s(k) = -\alpha\mathrm{sgn}[\bar{s}(k)],\qquad(3.25)$$

where α is a design parameter [19]. Thus, the combined control law, considering equivalent control u_{eq} and switching control u_s defined in (3.23) and (3.25) respectively, is obtained to be

$$u(k) = u_{eq}(k) + u_s(k).\qquad(3.26)$$

3.3.2.1 Modification in the Control Using Disturbance Estimator

The concept of disturbance estimation has proved to be effective in presence of slowly varying disturbances [1, 20]. The utilization of the disturbance estimator can make the bandwidth proportional to the rate of change in disturbance [20] and thus, disturbance estimation reduces the width of the DHOSM band. In this technique, an estimated term of disturbance vector, $\hat{d}_\xi(k)$, is added in the control input as

$$u(k) = u_{eq}(k) + u_s(k) - (\bar{c}\bar{b}_2)^{-1}(\bar{c}\hat{d}_\xi(k)).\qquad(3.27)$$

Since the disturbance is considered to be varying slowly, $d_\xi(k-1)$ would be a good estimate of $\hat{d}_\xi(k)$. In the problem under consideration, the effect of the delayed term of the disturbance vector $d_\xi(k)$ can be obtained as

$$\bar{c}d_\xi(k-1) = \bar{s}(k) - \bar{c}\left(\bar{A}_{21}\bar{x}(k-1) + \bar{A}_{22}\xi(k-1) + \bar{b}_2 u(k-1)\right).\qquad(3.28)$$

Since disturbance is bounded, the difference of two consecutive disturbance terms

$$\delta_\xi(k) = d_\xi(k) - d_\xi(k-1) \tag{3.29}$$

will also be bounded and there exists a known constant $\delta_{\xi max}$ as the maximum difference of two consecutive disturbance terms, i.e.

$$||\delta_\xi(k)|| \le \delta_{\xi max}. \tag{3.30}$$

The modified control input considering disturbance estimation technique is obtained to be

$$u(k) = u_{eq}(k) + u_s(k) - (\bar{c}\bar{b}_2)^{-1}(\bar{c}d_\xi(k-1)). \tag{3.31}$$

The stability of the discrete-time system under consideration with the proposed control law (3.31) is examined and is presented in the following section.

3.4 Stability Analysis

On using the modified control input (3.31) in the system (3.12), the resulting closed-loop system is obtained to be

$$\bar{x}(k+1) = \bar{A}_{11}\bar{x}(k) + \bar{A}_{12}\xi(k) \tag{3.32}$$

$$\xi(k+1) = A_c\xi(k) - \alpha\,\text{sgn}[\bar{s}(k)]\bar{b}_2 + \delta_\xi(k) \tag{3.33}$$

where $A_c = \begin{bmatrix} 0 & 1 & \cdots & 0 & 0 \\ 0 & 0 & \cdots & 1 & 0 \\ \vdots & \vdots & \ddots & \vdots & \vdots \\ 0 & 0 & \cdots & 0 & 1 \\ -\bar{a}_{n-r+1}+f_1 & -\bar{a}_{n-r+1}+f_2 & \cdots & -\bar{a}_{n-1}+f_{r-1} & -\bar{a}_n+f_r \end{bmatrix}.$

Lemma 3.1 *[21, Lemma 2] For a scalar dynamical system $z(k+1) = z(k) + g(k) - \epsilon\,\text{sgn}(z(k))$, suppose there exist two positive parameters γ and ϵ such that $0 \le |g(k)| < \gamma < \epsilon$, then the state z converges to the range confined by $|z| \le \epsilon + \gamma < 2\epsilon$.*

Theorem 3.2 *On using the control law (3.31) with parameter $\alpha > \dfrac{sup\left\|(\bar{c}A_c-\bar{c})\xi(k)+\bar{c}\delta_\xi(k)\right\|}{\bar{c}b_2}$ in the discrete-time system (3.12), ξ, $\bar{x}$ and $\underline{x}$ are ultimately bounded. Furthermore, the higher order sliding function is bounded by*

$$||\bar{s}|| \le 2\alpha\bar{c}\bar{b}_2. \tag{3.34}$$

Proof Consider (3.33),

$$\|\xi(k+1)\| = \left\|A_c\xi(k) - \alpha\mathrm{sgn}[\bar{s}(k)]\bar{b}_2 + \delta_\xi(k)\right\|$$
$$\leq \|A_c\|\,\|\xi(k)\| + \alpha\,\|\mathrm{sgn}[\bar{s}(k)]\|\,\|\bar{b}_2\| + \|\delta_\xi(k)\|$$

As $\|\mathrm{sgn}[\bar{s}(k)]\| = 1$, $\|\bar{b}_2\| = 1$, and $\|\delta_\xi(k)\| \leq \delta_{\xi max}$.

$$\|\xi(n)\| \leq \|A_c\|^n\,\|\xi(0)\| + \sum_{i=1}^{n-1}\|A_c\|^i\left(\alpha + \delta_{\xi max}\right),$$

$$\|\xi(n)\| \leq \|A_c\|^n\,\|\xi(0)\| + \left(\alpha + \delta_{\xi max}\right)\frac{\left[1 - \|A_c\|^n\right]}{\left[1 - \|A_c\|\right]}.$$

As the control input is designed such that spectral radius of A_c is less than one, ξ converges to a band bounded by

$$\|\xi\| \leq \frac{\alpha + \delta_{\xi max}}{\left[1 - \|A_c\|\right]} = \beta. \tag{3.35}$$

From (3.32),   .

$$\|\bar{x}(k+1)\| = \left\|\bar{A}_{11}\bar{x}(k) + \bar{A}_{12}\xi(k)\right\|,$$
$$\leq \left\|\bar{A}_{11}\right\|\,\|\bar{x}(k)\| + \left\|\bar{A}_{12}\right\|\,\|\xi(k)\|.$$

Thus,

$$\|\bar{x}(n)\| \leq \left\|\bar{A}_{11}\right\|^n\,\|\bar{x}(0)\| + \left(\left\|\bar{A}_{12}\right\|\,\|\xi\|\right)\sum_{i=0}^{n-1}\left\|\bar{A}_{11}\right\|^i,$$

$$\|\bar{x}(n)\| \leq \left\|\bar{A}_{11}\right\|^n\,\|\bar{x}(0)\| + \left(\left\|\bar{A}_{12}\right\|\beta\right)\frac{\left[1 - \left\|\bar{A}_{11}\right\|^n\right]}{\left[1 - \left\|\bar{A}_{11}\right\|\right]}.$$

As parameter c is chosen such that spectral radius of $\bar{A}_{11}$ is less than one, at steady state

$$\|\bar{x}\| \leq \left\|\bar{A}_{12}\right\|\beta\left[1 - \left\|\bar{A}_{11}\right\|\right]^{-1}. \tag{3.36}$$

On using state transformation (3.8), we have

$$\underline{x}(k) = G_2^{-1}\xi(k) - G_2^{-1}G_1\bar{x}(k),$$
$$\left\|\underline{x}(k)\right\| \leq \left\|G_2^{-1}\right\|\,\|\xi(k)\| + \left\|G_2^{-1}G_1\right\|\,\|\bar{x}(k)\|.$$

Thus, at steady state, $\underline{x}$ converges in a band

$$\|\underline{x}\| \le \|G_2^{-1}\|\beta + \|G_2^{-1}G_1\|\|\bar{A}_{12}\|\beta\left[1 - \|\bar{A}_{11}\|\right]^{-1}. \tag{3.37}$$

On using (3.3) and (3.33),

$$\bar{s}[\xi(k+1)] = \bar{s}[\xi(k)] + [\bar{c}A_c - \bar{c}]\xi(k) - \alpha\bar{c}\bar{b}_2\mathrm{sgn}[\bar{s}(k)] + \bar{c}\delta_\xi(k). \tag{3.38}$$

On using Lemma 3.1, it can be proved that if α is chosen such that $\alpha > \sup\frac{\|(\bar{c}A_c - \bar{c})\xi(k) + \bar{c}\delta_\xi(k)\|}{\bar{c}\bar{b}_2}$, then in steady-state value, $\|\bar{s}\| \le 2\alpha\bar{c}\bar{b}_2$. As it has been proved that ξ is bounded, s and its higher order terms are also bounded, and thus, DHOSM is established. This completes the proof.

3.5 Numerical and Experimental Validation

To validate the proposed control algorithm, the designed control law is numerically simulated and experimentally implemented on a rectilinear plant. The description of the setup of the rectilinear plant is given in Chap. 3. However in this chapter, unlike the Chap. 2, three serially connected mass–spring–damper are used in the experiment. The free body diagram of the system is shown in Fig. 3.1. This setup can be represented by a sixth-order discrete-time model (3.1) with parameters

$$A = \begin{bmatrix} 0 & 1 & 0 & 0 & 0 & 0 \\ 0 & 0 & 1 & 0 & 0 & 0 \\ 0 & 0 & 0 & 1 & 0 & 0 \\ 0 & 0 & 0 & 0 & 1 & 0 \\ 0 & 0 & 0 & 0 & 0 & 1 \\ -0.67 & 0.91 & -1.10 & 1.23 & -1.14 & 1.0157 \end{bmatrix} \tag{3.39}$$

and

$$b = \begin{bmatrix} 0 & 0 & 0 & 0 & 0 & 1 \end{bmatrix}^T. \tag{3.40}$$

It is assumed that among six states, only four states x_1, x_2, x_3, and x_4 are measurable. The parameter c in (3.2) is chosen such that the sliding function has relative degree $r = 3$ and has roots at points 0.1, 0.2 and 0.3. The k_f in (3.17) is chosen to be 15 samples.

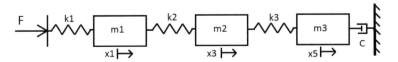

Fig. 3.1 Free body diagram of mass–spring system

3.5.1 Simulation Results

The numerical model of rectilinear plant is simulated in MATLAB to validate the proposed control algorithm . The initial condition is chosen to be $x(0) = \begin{bmatrix} 10 & 10 & 10 & 10 & 10 & 10 \end{bmatrix}^T$. The simulation results are presented in Fig. 3.2.

3.5.2 Experimental Results

For the experimental validation, the control law (3.31) is implemented to setup through MATLAB Simulink Real Time Windows Target. The experimental results of the proposed control algorithm are presented in Fig. 3.3. As shown in Fig. 3.3, on using optimal DHOSM algorithm, the higher order sliding function gradually decreases and becomes stable in the steady state. It can also be noticed that the higher order sliding function reaches the vicinity of higher order sliding surface at the a priori fixed sample $k_f = 15$. Subsequently, as shown in the inset, the higher order sliding function crosses the higher order sliding surface due to switching term. The optimal control input, shown in Fig. 3.3, is also bounded and becomes negligible once the higher order sliding function is stable. Figure 3.3 shows that the states of the system are bounded and demonstrates that the existence of DHOSM guarantees the boundedness of states.

Further, to examine the robustness of the proposed control, the external matched disturbance

$$d(k) = \begin{bmatrix} 0 \\ 0 \\ 0 \\ 0 \\ 0 \\ \sin(k/2) + \sin(k/100) \end{bmatrix} \tag{3.41}$$

is added to the system through disturbance motor. As shown in Fig. 3.4, the proposed control law rejects the effect of disturbance as $s(k)$ becomes less than the magnitude of disturbance $d(k-1)$. Thus, the proposed control algorithm has good disturbance rejection capability.

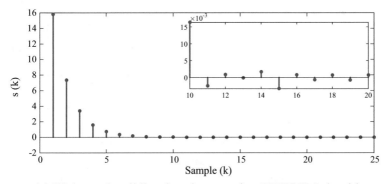

(a) Higher order sliding function on using ODHSOM algorithm.

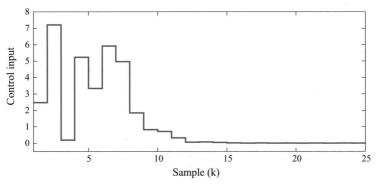

(b) Control input in ODHOSM algorithm.

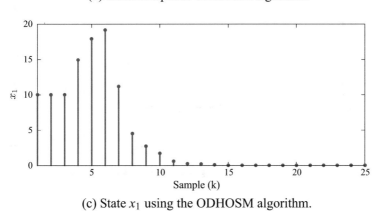

(c) State x_1 using the ODHOSM algorithm.

Fig. 3.2 Simulation results

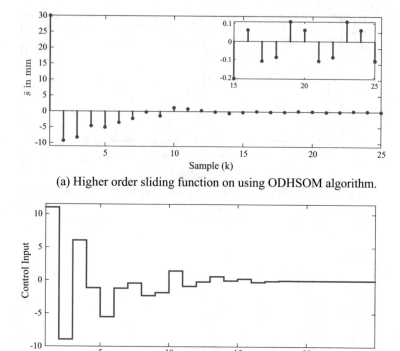

(a) Higher order sliding function on using ODHSOM algorithm.

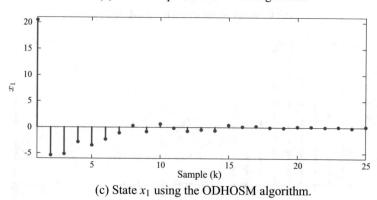

(b) Control input in ODHOSM algorithm.

(c) State x_1 using the ODHOSM algorithm.

Fig. 3.3 Experimental results

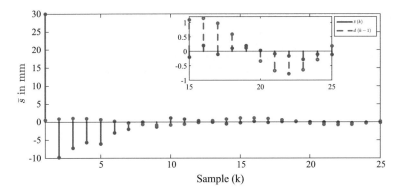

Fig. 3.4 Higher order sliding function on using ODHSOM algorithm in the presence of disturbance

3.6 Conclusion

In this chapter, an optimal discrete-time higher order sliding mode control is designed for an uncertain system with partial state information. An optimal higher order sliding surface is designed such that a quadratic cost function is minimized over a finite time period with a fixed final state. The corresponding control input guarantees that DHOSM takes place in a band. Further, the control algorithm is modified using disturbance estimation technique to reduce the DHSOM band. The system states, sliding function and its higher differences are proved to be bounded in the steady state. The proposed scheme is in experimentally validated on a rectilinear plant.

References

1. S. Janardhanan, V. Kariwala, IEEE Trans. Autom. Control **53**(1), 367 (2008)
2. C. Edwards, S. Spurgeon, *Sliding Mode Control: Theory and Applications* (CRC Press, 1998)
3. R. Galvn-Guerra, L. Fridman, IET Control Theor. Appl. **9**(4), 563 (2015)
4. P. Ignaciuk, A. Bartoszewicz, IEEE Trans. Autom. Control **55**(1), 269 (2010)
5. S. Y, S. J., ASME. J. Dyn. Sys., Meas. **139**(1), 014501 (2016)
6. F.J. Bejarano, L.M. Fridman, A.S. Poznyak, IEEE Trans. Autom. Control **54**(11), 2611 (2009)
7. M. Basin, P. Rodriguez-Ramirez, A. Ferrara, D. Calderon-Alvarez, J. Frankl. Inst. **349**(4), 1350 (2012). Special Issue on Optimal Sliding Mode Algorithms for Dynamic Systems
8. A. Bartoszewicz, A. Nowacka-Leverton, IEEE Trans. Autom. Control **55**(8), 1928 (2010)
9. S. Laghrouche, F. Plestan, A. Glumineau, Automatica **43**(3), 531 (2007)
10. M. Das, C. Mahanta, ISA Trans. **53**(6), 1807 (2014)
11. F. Dinuzzo, A. Ferrara, IEEE Trans. Autom. Control **54**(9), 2126 (2009)
12. N. Sun, Y. Niu, B. Chen, Optim. Control Appl. Methods **35**(4), 468 (2014)
13. P. Ignaciuk, A. Bartoszewicz, Int. J. Robust Nonlinear Control **19**(4), 442 (2009)
14. P. Ignaciuk, IEEE Trans. Autom. Control **58**(8), 2108 (2013)
15. M. Fink, J. Stecha, Kybernetika **07**(6), 467 (1971)
16. R.P. Agarwal, *Difference Equations and Inequalities: Theory, Methods, and Applications*. CRC Pure and Applied Mathematics (CRC Press, 2000)

17. A. Levant, Int. J. Control **58**(6), 1247 (1993)
18. D.S. Naidu, *Optimal Control Systems* (CRC Press Inc., Boca Raton, FL, USA, 2002)
19. B. Wang, X. Yu, X. Li, IEEE Trans. Ind. Electr. **55**(11), 4055 (2008)
20. S. Qu, X. Xia, J. Zhang, IEEE Trans. Ind. Electr. **61**(7), 3502 (2014)
21. X. Yu, B. Wang, Z. Galias, G. Chen, IEEE Trans. Autom. Control **53**(6), 1563 (2008)

Chapter 4
Discrete-Time Higher Order Sliding Mode Control of Unmatched Uncertain Systems

4.1 Introduction

Any model of a practical system would most probably have uncertainties due to unmodelled dynamics, external disturbance and other effects like friction. These uncertainties may not satisfy the so-called matching condition wherein the uncertainties affect the system through input channel. Therefore, these uncertainties cannot be eliminated through an equivalent control based SMC [1]. Unless these uncertainties are considered at the design stage, they may lead to unsatisfactory performance. Significant research in continuous-time sliding mode has been attempted to control unmatched uncertainties using first-order SMC [2–5] and higher order SMC [6]. In the case of discrete-time systems, some efforts have been made to design an integral discrete-time SMC to counter unmatched uncertainties [7–9].

In Chap. 3, a discrete-time higher order sliding mode control algorithm has been presented for an LTI system considering matched uncertainties. However, the control algorithms developed for the matched uncertainties may not be effective to control systems with unmatched uncertainties. Moreover, due to the presences of unmatched uncertainty in the system, the higher order sliding surface contains the disturbance terms. Since, the information of disturbance is not available, the control input designed in Chap. 3 cannot be applied for the systems with unmatched uncertainty. Further, in discrete-time sliding mode, disturbance estimation technique is used to minimize the effect of disturbance [10, 11]. However, the existing disturbance estimate techniques [10, 11] utilize only one past disturbance term and therefore, the disturbance signal cannot be effectively estimated. A technique is required which uses more than one past disturbance terms for better estimation of disturbances.

With these motivations, a discrete-time higher order sliding mode control is designed for LTI system in the presence of unmatched uncertainties is presented in this chapter. The major contributions of this paper are the following.

© Springer Nature Switzerland AG 2019
N. K. Sharma and J. Sivaramakrishnan, *Discrete-Time Higher Order Sliding Mode*,
https://doi.org/10.1007/978-3-030-00172-8_4

1. A discrete-time higher order sliding mode control is designed for an uncertain LTI system considering the presence of unmatched uncertainties. The unmatched uncertainties are considered at the design stage so that the control input effectively control the system in spite of the presence of unmatched uncertainties.
2. Due to the presence of unmatched uncertainty in the system, the DHOSM control requires the information of future uncertainty. For this purpose, a disturbance forecast technique is utilized to estimate the future disturbance terms. These estimated disturbance terms are used to modify the control input such that the closed-loop system is robust against disturbance.
3. The disturbance forecast technique utilizes a weighted moving average method. It considers more than one disturbance terms to estimate the current disturbance. Therefore, this technique is better than the existing disturbance estimation technique [11] which utilizes only one past disturbance term.
4. A series of experiments are conducted on an electromechanical rectilinear plant to validate and compare the performance of the designed controller with the existing controllers. In these experiments, various operational situations of industrial electromechanical systems are considered to realize the effect of unmatched uncertainty and performance of the presented control algorithm.

This should be noted that in any practical system, the effect of unmatched uncertainty cannot be negated. However, the effect can be minimized in the presence of a good estimate of the unmatched disturbance component. The paper approaches the control problem based on this logic.

The chapter is organized as follows. After the introduction in Sect. 4.1, the problem is formulated in Sect. 4.2. The formulated problem is solved by designing a DHOSM controller for system with unmatched uncertainties in Sect. 4.3. However, the designed control law requires the information of future uncertainties. Therefore, a technique is presented in Sect. 4.4 to forecast the unmatched uncertainties. Section 4.5 proves the stability of the closed-loop system. The simulation and experimental results of the designed control law are presented in Sect. 4.6. Finally, the chapter is concluded in Sect. 4.7.

4.2 Problem Formulation

Consider a discrete-time representation of an uncertain LTI system in controllable canonical form,

$$x(k + 1) = Ax(k) + bu(k) + d_x(k), \tag{4.1}$$

where $x \in \mathbb{R}^n$, $u \in \mathbb{R}$, and $d_x(k) \in \mathbb{R}^n$ are the state vector, control input and dis-
turbance vector, respectively. The matrix $A = \begin{bmatrix} 0 & 1 & 0 & \cdots & 0 \\ 0 & 0 & 1 & \cdots & 0 \\ \vdots & \vdots & \vdots & \ddots & \vdots \\ -a_1 & -a_2 & -a_3 & \cdots & -a_n \end{bmatrix}$, $b = \begin{bmatrix} 0 \\ 0 \\ \vdots \\ 1 \end{bmatrix}$

and $d_x = \begin{bmatrix} d_{x,1} \\ d_{x,2} \\ \vdots \\ d_{x,n} \end{bmatrix}$.

Unmatched Uncertainty [12]: An uncertain system (4.1) is said to have unmatched
uncertainty if the condition $d_x(k) \in \text{Range(B)}$ is not satisfied. In physical terms, this
assumption means that the disturbance affecting the system may also enter through
other than the input channel.

Assumption 5 The disturbance vector $d_x(k)$ represents the unmodelled dynamics
and the external disturbance effect in the system. It is unknown but norm bounded
by a known constant [11].

Assumption 6 The disturbance vector $d_x(k)$ is smooth. The smoothness assumption
made on the disturbance is to ensure that the disturbance bandwidth is sufficiently
lower than the controller bandwidth [9]. With the advent of high frequency electronics
components, the modern controllers have very high sampling rate and as a result, the
Nyquist frequency is very high. The disturbance signals which have frequency lower
than the Nyquist frequency are considered as smooth disturbances [13]. Since, the
Nyquist frequency is already very high and the practical external disturbance signals
are generally band limited, majority of the practical disturbance signals fall into the
category of smooth disturbance. Therefore, this assumption is not very restrictive for
the practical systems and frequently used in the literature [9].

Consider a sliding function defined as

$$s(k) = cx(k), \tag{4.2}$$

where the vector $c = [c_1, c_2, \ldots, c_{n-r-1}, c_{n-r}, 1, 0, \ldots, 0] \in \mathbb{R}^n$ contains coefficients such that sliding function $s \in \mathbb{R}$ has a relative degree r with respect to the
input u [14, p. 139]. As defined in (3.3) in Sect. 3.2, a higher order sliding function
is

$$\bar{s}(k) = \bar{c}\xi(k), \tag{4.3}$$

where

$$\xi(k) = \begin{bmatrix} s(k) \\ s(k+1) \\ \cdots \\ s(k+r-1) \end{bmatrix} \in \mathbb{R}^r, \tag{4.4}$$

and the vector $\bar{c} = [\bar{c}_1, \bar{c}_2, \bar{c}_3, \cdots, \bar{c}_{r-1}, 1] \in \mathbb{R}^r$. On using (4.2) in (4.4), we get

$$\xi(k) = \begin{bmatrix} cx(k) \\ cx(k+1) \\ \cdots \\ cx(k+r-1) \end{bmatrix} \tag{4.5}$$

Considering the system dynamics in (4.1) and the fact that $cb = cAb = cA^2b = \cdots = cA^{r-2}b = 0$, ξ can be derived to be

$$\xi(k) = Jx(k) + d_\xi(k), \tag{4.6}$$

where

$$J = \begin{bmatrix} c \\ cA \\ cA^2 \\ \vdots \\ cA^{r-1} \end{bmatrix}, \tag{4.7}$$

and

$$d_\xi(k) = \begin{bmatrix} 0 \\ cd_x(k) \\ cAd_x(k) + cd_x(k+1) \\ \vdots \\ \sum_{i=0}^{r-2} cA^i d_x(k+r-i-2) \end{bmatrix}. \tag{4.8}$$

The state x can be partitioned into two parts $\bar{x} = [x_1, x_2, \ldots, x_{n-r}]^T \in \mathbb{R}^{n-r}$ and $\underline{x} = [x_{n-r+1}, x_{n-r+2}, \cdots, x_n]^T \in \mathbb{R}^r$. With the partitioned states, the original system (4.1) is obtained to be

$$\begin{bmatrix} \bar{x}(k+1) \\ \underline{x}(k+1) \end{bmatrix} = \begin{bmatrix} A_{11} & A_{12} \\ A_{21} & A_{22} \end{bmatrix} \begin{bmatrix} \bar{x}(k) \\ \underline{x}(k) \end{bmatrix} + \begin{bmatrix} b_{\bar{x}} \\ b_{\underline{x}} \end{bmatrix} + \begin{bmatrix} d_{\bar{x}}(k) \\ d_{\underline{x}}(k) \end{bmatrix}. \tag{4.9}$$

In Chap. 3, the original system is transformed into a new form to design a DHOSM control with a vector

$$z(k) = [\bar{x}(k) \ \xi(k)]^T, \ z \in \mathbb{R}^n. \tag{4.10}$$

Similarly, using a transformation

$$z(k) = Gx(k) + d_z(k), \tag{4.11}$$

where G is as given in (3.9) and

$$d_z(k) = \begin{bmatrix} 0 \\ d_\xi(k) \end{bmatrix} \in \mathbb{R}^n, \tag{4.12}$$

we get

$$z(k) = \begin{bmatrix} I_{n-r} & 0 \\ G_1 & G_2 \end{bmatrix} \begin{bmatrix} \bar{x}(k) \\ \underline{x}(k) \end{bmatrix} + \begin{bmatrix} 0 \\ d_\xi(k) \end{bmatrix}. \tag{4.13}$$

From (4.11), $x(k)$ is expressed as

$$x(k) = G^{-1}(z(k) - d_z(k)). \tag{4.14}$$

On using (4.14) in (4.1), we get

$$z(k+1) = GAG^{-1}z(k) + Gbu(k) + Gd_x(k) - GAG^{-1}d_z(k) + d_z(k+1). \tag{4.15}$$

Thus, the original system (4.1) can be transformed into

$$z(k+1) = \bar{A}z(k) + \bar{b}u(k) + \bar{d}(k), \tag{4.16}$$

where

$$\bar{A} = GAG^{-1} = \begin{bmatrix} 0 & 1 & 0 & \cdots & 0 & 0 & 0 & \cdots & 0 & 0 \\ 0 & 0 & 1 & \cdots & 0 & 0 & 0 & \cdots & 0 & 0 \\ \vdots & \vdots & \ddots & \ddots & \vdots & \vdots & & \ddots & \vdots & \vdots \\ 0 & 0 & \cdots & 0 & 1 & 0 & 0 & \cdots & 0 & 0 \\ -c_1 & -c_2 & \cdots & -c_{n-r-1} & -c_{n-r} & 1 & 0 & \cdots & 0 & 0 \\ 0 & 0 & \cdots & 0 & 0 & 0 & 1 & 0 & \cdots & 0 \\ 0 & 0 & \cdots & 0 & 0 & 0 & 0 & 1 & \cdots & 0 \\ \vdots & \vdots & \ddots & \vdots & \vdots & \vdots & \vdots & \ddots & \ddots & \vdots \\ 0 & 0 & \cdots & 0 & 0 & 0 & 0 & \cdots & 0 & 1 \\ -\bar{a}_1 & -\bar{a}_2 & \cdots & -\bar{a}_{n-r-1} & -\bar{a}_{n-r} & -\bar{a}_{n-r+1} & -\bar{a}_{n-r+2} & \cdots & -\bar{a}_{n-1} & -\bar{a}_n \end{bmatrix}, \tag{4.17}$$

$$\bar{b} = Gb = \begin{bmatrix} 0 & 0 & \cdots & 0 & 1 \end{bmatrix}^T. \tag{4.18}$$

and

$$\bar{d}(k) = Gd_x(k) + d_z(k+1) - GAG^{-1}d_z(k). \tag{4.19}$$

On simplifying the individual terms

$$Gd_x(k) = G \begin{bmatrix} d_{x,1}(k) \\ d_{x,2}(k) \\ \vdots \\ d_{x,n-r}(k) \\ d_{x,n-r+1}(k) \\ d_{x,n-r+2}(k) \\ \vdots \\ d_{x,n}(k) \end{bmatrix} = \begin{bmatrix} d_{x,1}(k) \\ d_{x,2}(k) \\ \vdots \\ d_{x,n-r}(k) \\ cd_x(k) \\ cAd_x(k) \\ \vdots \\ cA^{r-1}d_x(k) \end{bmatrix}, \tag{4.20}$$

$$d_z(k+1) = \begin{bmatrix} 0 \\ \vdots \\ 0 \\ 0 \\ cd_x(k+1) \\ cAd_x(k+1) + cd_x(k+2) \\ \vdots \\ \sum_{i=0}^{r-2} cA^i d_x(k+r-i-1) \end{bmatrix}, \tag{4.21}$$

and

$$GAG^{-1}d_z(k) = \begin{bmatrix} 0 \\ \vdots \\ 0 \\ cd_x(k) \\ cAd_x(k) + cd_x(k+1) \\ \vdots \\ \sum_{i=0}^{r-2} cA^i d_x(k+r-i-2) \\ -\sum_{j=n-r+2}^{n} \bar{a}_j \sum_{i=0}^{r-2} cA^i d_x(k+r-i-2) \end{bmatrix} \tag{4.22}$$

On combining (4.20), (4.21) and (4.22),

$$\bar{d}(k) = Gd_x(k) + d_z(k+1) - GAG^{-1}d_z(k)$$

$$= \begin{bmatrix} d_{x,1}(k) \\ d_{x,2}(k) \\ \vdots \\ d_{x,n-r}(k) \\ 0 \\ 0 \\ 0 \\ \vdots \\ 0 \\ \sum_{i=0}^{r-1} cA^i d_x(k+r-i-1) - \sum_{j=n-r+2}^{n} \bar{a}_j \sum_{i=0}^{r-2} cA^i d_x(k+r-i-2) \end{bmatrix} \tag{4.23}$$

The $\bar{d}(k)$ in (4.23) can be represented as $\bar{d}(k) = \left[\bar{d}_1(k)\ \bar{d}_2(k)\right]^T$, where

$$\bar{d}_1(k) = \begin{bmatrix} d_{x,1}(k) \\ d_{x,2}(k) \\ \vdots \\ d_{x,n-r}(k) \end{bmatrix} \in \mathbb{R}^{(n-r)} \tag{4.24}$$

and

$$\bar{d}_2(k) = \begin{bmatrix} 0 \\ 0 \\ 0 \\ \vdots \\ 0 \\ \sum_{i=0}^{r-1} cA^i d_x(k+r-i-1) - \sum_{j=n-r+2}^{n} \bar{a}_j \sum_{i=0}^{r-2} cA^i d_x(k+r-i-2) \end{bmatrix} \in \mathbb{R}^{r} \tag{4.25}$$

The system (4.16) can be represented with the partitioned states as

$$\begin{bmatrix} \bar{x}(k+1) \\ \xi(k+1) \end{bmatrix} = \begin{bmatrix} \bar{A}_{11} & \bar{A}_{12} \\ \bar{A}_{21} & \bar{A}_{22} \end{bmatrix} \begin{bmatrix} \bar{x}(k) \\ \xi(k) \end{bmatrix} + \begin{bmatrix} \mathbf{0} \\ \bar{b}_2 \end{bmatrix} u(k) + \begin{bmatrix} \bar{d}_1(k) \\ \bar{d}_2(k) \end{bmatrix}, \tag{4.26}$$

where

$$\bar{A}_{11} = \begin{bmatrix} 0 & 1 & 0 & \cdots & 0 & 0 \\ 0 & 0 & 1 & \cdots & 0 & 0 \\ \vdots & \vdots & \vdots & \ddots & \vdots & \vdots \\ 0 & 0 & 0 & \cdots & 0 & 1 \\ -c_1 & -c_2 & -c_3 & \cdots & -c_{n-r-1} & -c_{n-r} \end{bmatrix},$$

$$\bar{A}_{12} = \begin{bmatrix} 0 & 0 & \cdots & 0 & 0 \\ 0 & 0 & \cdots & 0 & 0 \\ \vdots & \vdots & \ddots & \vdots & \vdots \\ 0 & 0 & \cdots & 0 & 0 \\ 1 & 0 & \cdots & 0 & 0 \end{bmatrix},$$

$$\bar{A}_{21} = \begin{bmatrix} 0 & 0 & \cdots & 0 \\ 0 & 0 & \cdots & 0 \\ \vdots & \vdots & \ddots & \vdots \\ 0 & 0 & \cdots & 0 \\ -\bar{a}_1 & -\bar{a}_2 & \cdots & -\bar{a}_{n-r} \end{bmatrix},$$

$$\bar{A}_{22} = \begin{bmatrix} 0 & 1 & 0 & \cdots & 0 & 0 \\ 0 & 0 & 1 & \cdots & 0 & 0 \\ \vdots & \vdots & \vdots & \ddots & \vdots & \vdots \\ 0 & 0 & 0 & \cdots & 0 & 1 \\ -\bar{a}_{n-r+1} & -\bar{a}_{n-r+2} & -\bar{a}_{n-r+3} & \cdots & -\bar{a}_{n-1} & -\bar{a}_n \end{bmatrix},$$

$b_2 = [0 \cdots 1]^T \in \mathbb{R}^r, \bar{d}_1 \in \mathbb{R}^{n-r}, \bar{d}_2 \in \mathbb{R}^r.$

The objective is to design a higher order sliding control law for a system (4.16) such that the discrete-time higher order sliding mode is achieved.

Remark 4.1 It should be noted that the problem is formulated in a similar manner as in Chap. 2, however due to the presence of unmatched uncertainty, the transformed model (4.26) contains the disturbance terms of future. Therefore, ξ cannot be computed directly and the existing control law, structured in Chap. 2, cannot control the system efficiently. Hence, a control law is required which can tackle the unmatched uncertainties present in the system.

4.3 Control Design

As discussed in Chap. 3, the DHOSM control input has two parts: equivalent control, u_{eq} and switching control, u_s.

4.3.1 Equivalent Control Design

Consider the subsystem of (4.26) in the absence of disturbance with the equivalent control part,

$$\xi(k+1) = \bar{A}_{21}\bar{x}(k) + \bar{A}_{22}\xi(k) + \bar{b}_2 u_{eq}(k). \tag{4.27}$$

The equivalent control part should ensure that

$$\bar{s}(k+1) = 0. \tag{4.28}$$

On using (4.27) in (4.28), we get

$$\bar{s}(k+1) = \bar{c}\left(\bar{A}_{21}\bar{x}(k) + \bar{A}_{22}\xi(k) + \bar{b}_2 u_{eq}(k)\right) = 0. \tag{4.29}$$

Thus, the equivalent control can be synthesized to be

$$u_{eq}(k) = -\left(\bar{c}\bar{b}_2\right)^{-1}\left(\bar{c}\bar{A}_{21}\bar{x}(k) + \bar{c}\bar{A}_{22}\xi(k)\right) \tag{4.30}$$

The computation of the equivalent control (4.30) requires the information of $\xi(k)$. The vector $\xi(k)$, defined in (4.6), contains $x(k)$ and $d_\xi(k)$. Although the state x is measurable, $d_\xi(k)$ contains the terms $d(k+j) \; \forall j = 0$ to $r - 2$, which are the disturbance terms of future and thus, the information of $d_\xi(k)$ is not available for the control design. Nevertheless, $d_\xi(k)$ and $\hat{\bar{d}}_2$ can be forecast using the weighted moving average method. The forecast value of $\xi(k)$ and $\bar{d}_2(k)$ is denoted by $\hat{\xi}(k)$ and $\hat{\bar{d}}_2$, respectively. The computation method of $\hat{\xi}(k)$ and $\hat{\bar{d}}_2$ is given in the next section.

4.3.2 Switching Control Design

To compensate the uncertainties in the system, the switching control part $u_s(k)$ is designed to be

$$u_s(k) = -\alpha \text{sgn}[\hat{\bar{s}}(k)], \tag{4.31}$$

where $\hat{\bar{s}}(k) = \bar{c}\hat{\xi}(k)$ and α is a design parameter [15].

Thus, the combined control law, considering the equivalent control u_{eq} and the switching control u_s, defined in (4.30) and (4.31) respectively, is obtained to be

$$u(k) = - (\bar{c}b_2)^{-1} \bar{c}\bar{A}_{21}\bar{x}(k) - (\bar{c}\bar{b})^{-1}\bar{c}\bar{A}_{22}\hat{\xi}(k) - (\bar{c}\bar{b})^{-1}\bar{c}\hat{\bar{d}}_2(k) - \alpha\text{sgn}(\hat{\bar{s}}(k)). \tag{4.32}$$

4.4 Disturbance Forecasting

Disturbance forecasting is a well-established concept and frequently used technique in systems with smooth disturbances [16–19]. There are many forecasting techniques, such as discrete Kalman filter, weighted moving average, double exponential smoothing, etc [20]. The disturbance terms are represented as time series structure and are utilized to forecast future values of disturbances. Since it is assumed that the disturbance is smooth, the forecasting technique can be utilized in this paper. A simple but efficient technique of weighted moving average method is used here. In this method, the weighted average of certain number of past disturbances is considered to forecast the disturbance of future [21]. For example, the disturbance $d_x(k+j)$ can be forecast using the last p terms as

$$d_x(k+j) = \frac{\phi_1 d_x(k+j-1) + \phi_2 d_x(k+j-2) + \cdots + \phi_p d_x(k+j-p)}{\sum_{i=1}^{p} \phi_i} \tag{4.33}$$

$$d_x(k+j) = \frac{\sum_{i=1}^{p} \phi_i d_x(k+j-i)}{\sum_{i=1}^{p} \phi_i}, \tag{4.34}$$

where ϕ_i's are the weight coefficients. Weighted moving average method assigns larger weights to more recent data points since they are more relevant than data points in the distant past. Therefore, weight coefficient associated with the recent disturbance should be greater than the previous one, i.e. the weight coefficients $\phi_i > \phi_{i-1} \; \forall \, i = 1$ to p. Thus, the weight coefficients in this paper are chosen to be

$$\phi_i = p - i + 1, \quad \forall i = 1 \text{ to } p, \tag{4.35}$$

i.e. $\phi_1 = p, \phi_2 = p - 1, ..., \phi_{p-1} = 2, \phi_p = 1$. Here, the weight coefficients are assigned linearly in the decreasing order. However, there can be other ways of choosing weights coefficients according to the problem specification. In the problem under consideration, the disturbance term $d_x(k - 1)$ can be computed using the system dynamics (4.1) and the past control input as

$$d_x(k - 1) = x(k) - Ax(k - 1) - bu(k - 1). \tag{4.36}$$

With the knowledge of $d_x(k - 1)$ and the other previous disturbance terms, the disturbance term $d_x(k)$ can be forecast using (4.34) as

$$\hat{d}_x(k) = \frac{\phi_1 d_x(k - 1) + \phi_2 d_x(k - 2) + \cdots + \phi_p d_x(k - p)}{\sum_{i=1}^{p} \phi_i}. \tag{4.37}$$

Similarly, $d_x(k + 1)$ can be forecast using the forecast value $\hat{d}_x(k)$, from (4.37), and the past values of disturbance as

$$\hat{d}_x(k + 1) = \frac{\phi_1 \hat{d}_x(k) + \phi_2 d_x(k - 1) + \cdots + \phi_p d_x(k + 1 - p)}{\sum_{i=1}^{p} \phi_i}. \tag{4.38}$$

Thus, a general equation to forecast $\hat{d}_x(k + j)$ can be obtained as

$$\hat{d}_x(k + j) = \frac{\sum_{i=1}^{j} \phi_i \hat{d}_x(k + j - i) + \sum_{i=j+1}^{p} \phi_i d_x(k + j - i)}{\sum_{i=1}^{p} \phi_i}. \tag{4.39}$$

Thus, $d_\xi(k)$, given in (4.8), can be forecast and its forecast value is

$$\hat{d}_\xi(k) = \begin{bmatrix} 0 \\ c\hat{d}_x(k) \\ cA\hat{d}_x(k) + c\hat{d}_x(k + 1) \\ \vdots \\ \sum_{i=0}^{r-2} cA^i \hat{d}_x(k + r - i - 2) \end{bmatrix}. \tag{4.40}$$

Since x is measurable, with the known $\hat{d}_\xi(k)$, $\hat{\xi}(k)$ can be obtained on using relation

$$\hat{\xi}(k) = Jx(k) + \hat{d}_\xi(k). \tag{4.41}$$

Similarly, $\bar{d}_2(k)$, given in (4.25), can also be forecast as

$$\hat{d}_2(k) = \begin{bmatrix} 0 \\ 0 \\ 0 \\ \vdots \\ 0 \\ \sum_{i=0}^{r-1} cA^i \hat{d}_x(k+r-i-1) - \sum_{j=n-r+2}^{n} \bar{a}_j \sum_{i=0}^{r-2} cA^i \hat{d}_x(k+r-i-2) \end{bmatrix} \tag{4.42}$$

Therefore, the modified combined control is

$$u(k) = -(\bar{c}b_2)^{-1}\bar{c}\bar{A}_{21}\bar{x}(k) - (\bar{c}b)^{-1}\bar{c}\bar{A}_{22}\hat{\xi}(k) - (\bar{c}b)^{-1}\bar{c}\hat{d}_2(k) - \alpha\,\mathrm{sgn}(\hat{s}(k)). \tag{4.43}$$

where

$$\hat{s}(k) = \bar{c}\hat{\xi}(k) \tag{4.44}$$

However, the forecasting is not perfect and susceptible to errors. Let the forecasting error in $\xi(k)$ is defined as

$$\tilde{\xi}(k) = \xi(k) - \hat{\xi}(k). \tag{4.45}$$

Similarly, the forecasting error $\tilde{d}_{\xi_2}$ and $\tilde{d}_2$ are defined as

$$\tilde{d}_{\xi_2}(k) = d_\xi(k) - \hat{d}_\xi(k), \tag{4.46}$$

and

$$\tilde{d}_2(k) = \bar{d}_2(k) - \hat{d}_2(k). \tag{4.47}$$

This relation can also be rewritten in the form

$$\hat{d}_2(k) = \bar{d}_2(k) - \tilde{d}_2(k). \tag{4.48}$$

It is to be noted that the moving average of bounded terms is always bounded. Since the disturbance is assumed to be bounded, its moving average $\hat{d}_x$ will also be bounded. Similarly, since $\hat{d}_\xi$ and $\hat{d}_2$, given in (4.40) and (4.42) respectively, are the linear combination of $\hat{d}_x$, so $\hat{d}_\xi$ and $\hat{d}_2$ will also be bounded. Therefore, there exist constants $\tilde{d}_{\xi max}$ and $\tilde{d}_{2max}$ such that

$$\|\tilde{d}_\xi\| \le \tilde{d}_{\xi max}. \tag{4.49}$$

and

$$||\tilde{\tilde{d}}_2|| \leq \tilde{\tilde{d}}_{2max}. \tag{4.50}$$

The effect of these forecasting errors on the overall closed-loop system stability is analysed in Sect. 4.5.

4.5 Stability Analysis

On using the control input (4.43) in (4.27), the closed-loop system is obtained to be

$$\xi(k+1) = \bar{A}_{21}\bar{x}(k) + \bar{A}_{22}\xi(k) + \bar{b}_2(-(\bar{c}b_2)^{-1}\bar{c}\bar{A}_{21}\bar{x}(k) - (\bar{c}\bar{b})^{-1}\bar{c}\bar{A}_{22}\hat{\xi}(k)$$
$$- (\bar{c}\bar{b})^{-1}\bar{c}\hat{\bar{d}}_2(k) - \alpha\mathrm{sgn}(\hat{\bar{s}}(k))) + \bar{d}_2(k). \tag{4.51}$$

On simplifying, we get

$$\xi(k+1) = \bar{A}_{22}\xi(k) - \bar{b}_2\left((\bar{c}\bar{b}_2)^{-1}\bar{c}\bar{A}_{22}\hat{\xi}(k) - b_2\alpha\mathrm{sgn}(\hat{\bar{s}}(k))\right) - \bar{b}_2(\bar{c}\bar{b}_2)^{-1}\bar{c}\hat{\bar{d}}_2(k) + \bar{d}_2(k). \tag{4.52}$$

On using the forecasting errors from (4.45) and (4.47) to (4.52), we get

$$\xi(k+1) = [\bar{A}_{22} - \bar{b}_2(\bar{c}\bar{b}_2)^{-1}\bar{c}\bar{A}_{22}]\xi(k) + [\bar{d}_2(k) - \bar{b}_2(\bar{c}\bar{b}_2)^{-1}\bar{c}\bar{d}_2(k)]$$
$$+ \bar{b}_2(\bar{c}\bar{b}_2)^{-1}\bar{c}\bar{A}_{22}\tilde{\xi}(k) + \bar{b}_2(\bar{c}\bar{b}_2)^{-1}\bar{c}\tilde{\bar{d}}_2(k) - b_2\alpha\mathrm{sgn}(\hat{\bar{s}}(k)). \tag{4.53}$$

It should be noticed that in $\xi(k)$, given by (4.6), the state $x(k)$ is measurable and the uncertainty is due to the disturbance part alone. On using (4.6) and (4.41) in (4.45), we get

$$\tilde{\xi}(k) = Jx(k) + d_\xi(k) - Jx(k) - \hat{d}_\xi(k) = d_\xi(k) - \hat{d}_\xi(k) = \tilde{d}_\xi(k). \tag{4.54}$$

Thus, the forecasting error in ξ is equal to the forecasting error of $\tilde{d}_\xi(k)$. Further, it should be noted that $\bar{d}_2 \in span\,(b_2)$. Thus

$$[I - \bar{b}_2(\bar{c}\bar{b}_2)^{-1}\bar{c}]\bar{d}_2(k) = 0. \tag{4.55}$$

On using the relations (4.54) and (4.55) in (4.53), we get

$$\xi(k+1) = [\bar{A}_{22} - \bar{b}_2(\bar{c}\bar{b}_2)^{-1}\bar{c}\bar{A}_{22}]\xi(k) + \bar{b}_2(\bar{c}\bar{b}_2)^{-1}\bar{c}\bar{A}_{22}\tilde{d}_\xi(k) + \bar{b}_2(\bar{c}\bar{b}_2)^{-1}\bar{c}\tilde{\bar{d}}_2(k)$$
$$- b_2\alpha\mathrm{sgn}(\hat{\bar{s}}(k)), \tag{4.56}$$

which can be simplified to

$$\xi(k+1) = \Phi\xi(k) + \delta_\xi(k) - b_2\alpha\mathrm{sgn}(\hat{\bar{s}}(k)). \tag{4.57}$$

where

$$\Phi = [\bar{A}_{22} - \bar{B}_2(\bar{c}\bar{B}_2)^{-1}\bar{c}\bar{A}_{22}],$$

and

$$\delta_\xi(k) = \bar{B}_2(\bar{c}\bar{B}_2)^{-1}\bar{c}\bar{A}_{22}\tilde{d}_\xi(k) + \bar{B}_2(\bar{c}\bar{B}_2)^{-1}\bar{c}\tilde{\bar{d}}_2(k).$$

Since $\tilde{d}_\xi$ and $\tilde{\bar{d}}_2$ are bounded with their bounds given in (4.49) and (4.50), and δ_ξ is a linear combination of $\tilde{d}_\xi$ and $\tilde{\bar{d}}_2$, there exists a constant $\delta_{\xi max}$ such that

$$||\delta_\xi|| \leq \delta_{\xi max}. \tag{4.58}$$

Here δ_ξ depends upon the error in the disturbance forecast. Thus, better the accuracy of forecast algorithm, smaller the value of $\delta_{\xi max}$. The steady-state bounds of states, sliding function and higher order sliding function are presented in Theorem 4.1.

Theorem 4.1 *On using the control law (4.43) in an uncertain system (4.1) in the presence of unmatched uncertainty, states $\bar{x}$, $\underline{x}$, and ξ are ultimately bounded and discrete-time higher order sliding mode is established.*

Proof Consider (4.57),

$$||\xi(k+1)|| = \left\| \Phi\xi(k) + \delta_\xi(k) + b_2\alpha\,\text{sgn}(\hat{\bar{s}}(k)) \right\|,$$

$$\leq ||\Phi||\,||\xi(k)|| + \left\| \delta_\xi(k) \right\| + \left\| b_2\alpha\,\text{sgn}(\hat{\bar{s}}(k)) \right\|.$$

$$||\xi(n)|| \leq ||\Phi||^n\,||\xi(0)|| + \sum_{i=1}^{n-1} ||\Phi||^i\,(\delta_{\xi max} + \alpha),$$

$$||\xi(n)|| \leq ||\Phi||^n\,||\xi(0)|| + (\delta_{\xi max} + \alpha)\frac{[1 - ||\Phi||^n]}{[1 - ||\Phi||]}.$$

The vector $\bar{c}$ is chosen such that the spectral radius of Φ is less than one. Therefore, ξ converges to a band bounded by

$$||\xi|| \leq \frac{\delta_{\xi max} + \alpha}{[1 - ||\Phi||]} = \beta. \tag{4.59}$$

From (4.26),

$$\|\bar{x}(k+1)\| = \left\| \bar{A}_{11}\bar{x}(k) + \bar{A}_{12}\xi(k) + \bar{d}_1(k) \right\|,$$
$$\leq \left\| \bar{A}_{11} \right\| \|\bar{x}(k)\| + \left\| \bar{A}_{12} \right\| \|\xi(k)\| + \left\| \bar{d}_1(k) \right\|.$$

Thus, the nth sample

$$\|\bar{x}(n)\| \leq \|\bar{A}_{11}\|^n \|\bar{x}(0)\| + \sum_{i=0}^{n-1} \|\bar{A}_{11}\|^i \|\bar{A}_{12}\| \|\xi(i)\| + \sum_{i=0}^{n-1} \|\bar{A}_{11}\|^i \|\bar{d}_1(i)\|, \quad (4.60)$$

Since $\|\bar{d}_1\| \leq \bar{d}_{1max}$ and $\|\xi\| \leq \beta$.

$$\|\bar{x}(n)\| \leq \|\bar{A}_{11}\|^n \|\bar{x}(0)\| + \|\bar{A}_{12}\|\beta \sum_{i=0}^{n-1} \|\bar{A}_{11}\|^i + \bar{d}_{1max} \sum_{i=0}^{n-1} \|\bar{A}_{11}\|^i,$$

$$\leq \|\bar{A}_{11}\|^n \|\bar{x}(0)\| + (\|\bar{A}_{12}\|\beta + \bar{d}_{1max}) \sum_{i=0}^{n-1} \|\bar{A}_{11}\|^i.$$

$$\leq \|\bar{A}_{11}\|^n \|\bar{x}(0)\| + (\|\bar{A}_{12}\|\beta + \bar{d}_{1max}) \frac{\left[1 - \|\bar{A}_{11}\|^{n-1} \right]}{\left[1 - \|\bar{A}_{11}\| \right]}.$$

As parameter c is chosen such that spectral radius of $\bar{A}_{11}$ is less than one, at steady state

$$\|\bar{x}\| \leq \frac{(\|\bar{A}_{12}\|\beta + \bar{d}_{1max})}{\left[1 - \|\bar{A}_{11}\| \right]}. \quad (4.61)$$

From relation (4.13), we have

$$\underline{x}(k) = G_2^{-1}\xi(k) - G_2^{-1}G_1\bar{x}(k) - G_2^{-1}d_\xi(k),$$
$$\|\underline{x}(k)\| \leq \|G_2^{-1}\| \|\xi(k)\| + \|G_2^{-1}G_1\| \|\bar{x}(k)\| + \|G_2^{-1}G_1\| \|d_\xi(k)\|.$$

At steady state, $\underline{x}$ converges in a band

$$\|\underline{x}\| \leq \|G_2^{-1}\| \beta + \|G_2^{-1}G_1\| \left[(\|\bar{A}_{12}\|\beta + \bar{d}_{1max}) \left[1 - \|\bar{A}_{11}\| \right]^{-1} + \|d_\xi(k)\| \right]. \quad (4.62)$$

From (4.44), $\hat{\bar{s}}(k+1)$ can be written in the form

$$\hat{\bar{s}}(k+1) = \bar{c}\hat{\bar{\xi}}(k+1) + \hat{\bar{s}}(k) - \bar{c}\hat{\bar{\xi}}(k). \quad (4.63)$$

On using error dynamics (4.54) in (4.63), we get

$$\hat{\bar{s}}(k+1) = \bar{c}\left(\xi(k+1) - \tilde{\xi}(k+1) \right) + \hat{\bar{s}}(k) - \bar{c}\left(\xi(k) - \tilde{\xi}(k) \right). \quad (4.64)$$

Equation (4.64) can be further simplified using (4.57),

$$\hat{\bar{s}}(k+1) = \hat{\bar{s}}(k) + [\bar{c}\Phi - \bar{c}]\xi(k) + \bar{c}\delta_\xi(k) - \bar{c}\bar{B}_2\alpha\mathrm{sgn}(\hat{\bar{s}}(k)) - \bar{c}\tilde{\xi}(k+1) + \bar{c}\tilde{\xi}(k). \tag{4.65}$$

On using (4.54), (4.65) is obtained to be

$$\hat{\bar{s}}(k+1) = \hat{\bar{s}}(k) + [\bar{c}\Phi - \bar{c}]\xi(k) + \delta_{\bar{s}}(k) - \bar{c}\bar{B}_2\alpha\mathrm{sgn}(\hat{\bar{s}}(k)), \tag{4.66}$$

where

$$\delta_{\bar{s}}(k) = \bar{c}\delta_\xi(k) - \bar{c}\tilde{d}_\xi(k+1) + \bar{c}\tilde{d}_\xi(k), \tag{4.67}$$

$$||\delta_{\bar{s}}|| = ||\bar{c}\delta_\xi(k) - \bar{c}\tilde{d}_\xi(k+1) + \bar{c}\tilde{d}_\xi(k)||,$$

$$\leq ||\bar{c}\delta_\xi(k)|| + ||\bar{c}\tilde{d}_\xi(k+1)|| + ||\bar{c}\tilde{d}_\xi(k)||.$$

Since $\tilde{d}_\xi$ and δ_ξ are bounded with their bounds given in (4.49) and (4.58), respectively. On using these bound,

$$||\delta_{\bar{s}}(k)|| \leq ||\bar{c}||[2\tilde{d}_{\xi max} + \delta_{\xi max}]. \tag{4.68}$$

Thus, there exists a bound

$$\delta_{\bar{s},max} = ||\bar{c}||[2\tilde{d}_{\xi max} + \delta_{\xi max}], \tag{4.69}$$

such that

$$||\delta_{\bar{s}}(k)|| \leq \delta_{\bar{s},max}. \tag{4.70}$$

The bound $\delta_{\bar{s},max}$ depends upon the error in the disturbance forecast. Thus, better the accuracy of forecast algorithm, smaller the value of $\delta_{\bar{s},max}$.

On using Lemma 3.1, it can be proved that if α is chosen to be $\alpha > \frac{[\bar{c}\Phi - \bar{c}]\beta + \delta_{\bar{s},max}}{\bar{c}\bar{B}_2}$ then in steady state $||\bar{s}|| \leq 2\alpha\bar{c}\bar{B}_2$. As it is proved that ξ is bounded, s and its higher order terms are also bounded, and thus, DHOSM is established. This completes the proof.

The control algorithm presented in the paper is experimentally validated on a rectilinear plant and the results are presented in Sect. 4.6.2.

4.6 Numerical and Experimental Validation

To validate the presented control algorithm, the designed control law is numerically simulated and experimentally implemented on a rectilinear plant. Since, the rectilinear plant is a mass–spring–damper system and is very susceptible to uncertainties, this system is the most appropriate setup for the study under consideration. The description of the rectilinear plant and its dynamics are given in Sect. 3.5 of Chap. 3.

4.6.1 Simulation Results

The numerical model of rectilinear plant is simulated in MATLAB with the initial condition $x(0) = \begin{bmatrix} 10 & 10 & 10 & 10 & 10 & 10 \end{bmatrix}^T$ and unmatched disturbance

$$d_x(k) = \begin{bmatrix} 0.2 \\ -0.2 \\ -0.1 \\ 0.1 \\ -0.1 \\ 0.1 \end{bmatrix} \sin(k/2) + \sin(k/100). \tag{4.71}$$

The parameter c in (4.2) is chosen such that the sliding function has relative degree $r = 3$ and has roots at points 0.1, 0.2 and 0.3. For disturbance forecasting, the last 5 disturbance terms are utilized using (4.39). The weight coefficients are found out using (4.35) as $\phi_1 = 5, \phi_2 = 4, \phi_3 = 3$, $\phi_4 = 2$, and $\phi_5 = 1$.

The simulation results are presented in Fig. 4.1. To show the boundedness of the closed-loop system, one of the states, x_3 is shown in Fig. 4.1a. It can be noticed that the state x_3 gradually starts approaching the origin from initial condition and then remains in a bound in steady state. The variation in the steady state is due to the external disturbance injected in the system. The control input, shown in Fig. 4.1b, starts decreasing as the system's state approaches steady state and then remains bounded.

The disturbance estimation technique utilized in this paper uses 5 past disturbance terms. On the other hand, disturbance estimation technique presented in [10, 11] uses only one past disturbance term. To show the superiority of the disturbance estimation technique utilized in this paper, the same numerical example is simulated with disturbance estimation technique utilized in [11]. The absolute sum of the state, $\sum_i |x(i)|$, and the control input, $\sum_i |u(i)|$, can quantify the total deviation from the origin and the total control efforts, respectively. The absolute sum of state using the presented disturbance estimation technique is found out to be 53.33. On the other hand, the absolute sum of state obtained using the disturbance estimation technique presented in [11] is found out to be 54.43. Similarly, the absolute sum of control input on using the presented disturbance estimation technique is found out to be 23.92. While, the absolute sum of control input obtained using the disturbance estimation technique presented in [11] is found out to be 54.43. Thus, the control input using the proposed technique is found to be lesser and smoother as compared to the control input using disturbance estimation technique presented in [11] . Hence, the disturbance estimation technique utilized in this paper exhibits better performance than the disturbance estimation technique presented in [11].

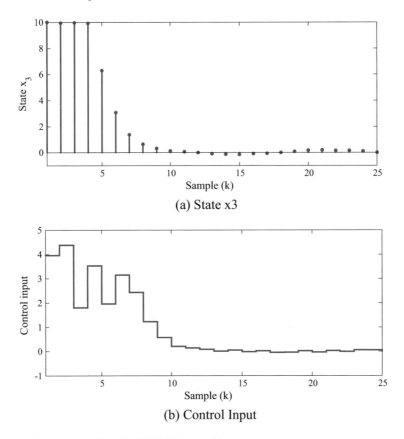

Fig. 4.1   Simulation results of the UDHOSM control

4.6.2   *Experimental Results*

A series of experiments are conducted for the implementation of the presented controllers on an electromechanical rectilinear plant. Three experiments are conducted for validation of the controller. Each experiment represents a situation which usually occurs during operation of an electromechanical system in an industry.

4.6.2.1   Experiment 1: External Disturbance Added Through Disturbance Motor

Usually, electromechanical systems in industries may encounter unmatched external disturbance. This situation is realized practically through connecting a disturbance motor to m_2 and applying an external disturbance. The disturbance signal $\sin(k/2) + \sin(k/100)$ contains mixed frequency signals which imitate disturbance

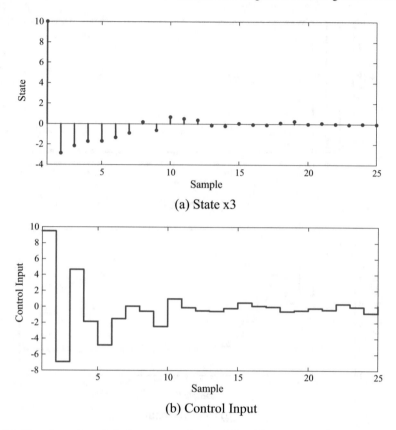

(a) State x3

(b) Control Input

Fig. 4.2 Experimental results in the presence of external disturbances introduced by the disturbance motor

in real scenario. The experimental results are depicted in Fig. 4.2. To verify the performance of the control algorithm, the state x_3 is plotted.

As shown in Fig. 4.2a, on using DHOSM control, the state x_3 of the system gradually decreases and becomes stable in the steady state. The control input, shown in Fig. 4.2b, is also bounded and becomes lesser once the higher order sliding function is stable. Thus, these figures demonstrate that the existence of DHOSM guarantees the boundedness of state.

4.6.2.2   Experiment 2: Uncertainties Due to Mass Variation

Generally, the industrial electromechanical systems are modelled considering the nominal mass. But, in real life scenario, their practical mass may vary from their nominal mass and thus, the unmodelled dynamics infiltrate in the system. To emulate this situation in the experimental setup, the mass of m_2 is reduced by 0.5 Kg and

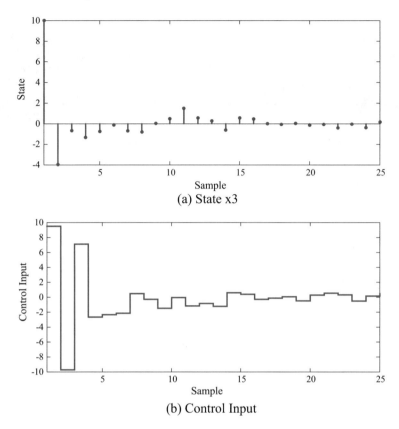

(a) State x3

(b) Control Input

Fig. 4.3 Experimental results in presence of external disturbances due to mass variation

the mass of m_3 is increased by 0.5 Kg. The control input is designed considering $m_2=m_3=2.59$ Kg, but practically the masses are $m_2 = 2.09$ Kg and $m_3 = 3.09$ Kg. This action introduces unmatched unmodelled dynamics in the system.

The experimental results in this scenario are depicted in Fig. 4.3. It can be seen in Fig. 4.3a that due the mass variation, the state x_3 shows more fluctuation than the previous experiment before becoming steady at around $k=15$. In the steady state, the mass behaves almost same as the previous experiment. Similarly, Fig. 4.3b shows that the required control efforts to stabilize the mass is more than the previous experiment.

4.6.2.3   Experiment 3: Uncertainties Due to Inclined Base

Primarily, the controllers of the electromechanical equipments are designed considering their base horizontal to the ground. However, occasionally, the base of the system may become incline during course of the operation. In this situation, some unmodelled dynamics due to gravity and other reasons are added in the system. As

Fig. 4.4 Experimental Setup of the rectilinear plant with inclined base

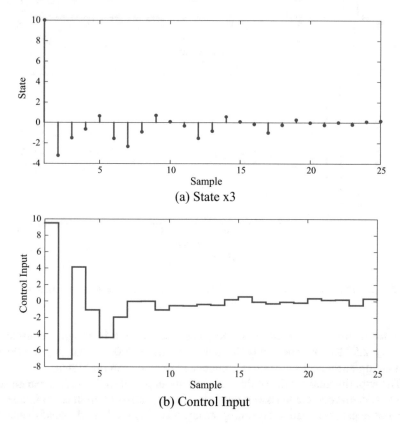

Fig. 4.5 Experimental results in presence of external disturbances due to mass variation

a result, the performance of the system is not as desired. To realize this situation practically, the base of the rectilinear plant has been tilted 30° from the ground, as shown in Fig. 4.4. The experimental results are given in the Fig. 4.5. It is observed that inclined base causes the larger overshoot in the negative direction of the mass and more unmatched uncertainties in the electromechanical plant. The control efforts are also maximum in the situation of the inclined plain.

4.7 Conclusion

In this chapter, a discrete-time higher order sliding mode controller is presented for an uncertain LTI system with unmatched uncertainties. The DHOSM control includes the forecast of future disturbance signal using the weighted moving average method. The system states, sliding function and its higher differences are analytically proved to be bounded in the steady state. The proposed scheme is experimentally validated on a rectilinear plant through conducting three experiments to imitate various practical situations. The proposed technique is found to be very effective in controlling systems with unmatched uncertainty.

References

1. S. Janardhanan, B. Bandyopadhyay, IEEE Trans. Autom. Control **51**(6), 1030 (2006)
2. Y.W. Liang, L.W. Ting, L.G. Lin, IEEE Trans. Ind. Electr. **59**(8), 3062 (2012)
3. J. Yang, S. Li, X. Yu, IEEE Trans. Ind. Electr. **60**(1), 160 (2013)
4. D. Ginoya, P.D. Shendge, S.B. Phadke, IEEE Trans. Ind. Electr. **61**(4), 1983 (2014)
5. F. Castanos, L. Fridman, IEEE Trans. Autom. Control **51**(5), 853 (2006)
6. A. Estrada, L. Fridman, Automatica **46**(11), 1916 (2010)
7. Z. Xi, T. Hesketh, IET Control Theor. Appl. **4**(5), 889 (2010)
8. Z. Xi, T. Hesketh, IET Control Theor. Appl. **4**(10), 2071 (2010)
9. K. Abidi, J.X. Xu, Y. Xinghuo, IEEE Trans. Autom. Control **52**(4), 709 (2007)
10. N.K. Sharma, S. Janardhanan, Int. J. Robust Nonlinear Control **27**(17), 4104 (2017). https://doi.org/10.1002/rnc.3785. Rnc. 3785
11. S. Qu, X. Xia, J. Zhang, IEEE Trans. Ind. Electr. **61**(7), 3502 (2014)
12. S. Janardhanan, V. Kariwala, IEEE Trans. Autom. Control **53**(1), 367 (2008)
13. B. Bandyopadhyay, S. Janardhanan, *Discrete-time Sliding Mode Control : A Multirate-Output Feedback Approach* (Ser. Lecture Notes in Control and Information Sciences, Vol. 323, Springer-Verlag, 2005)
14. A. Isidori, *Nonlinear Control Systems*, 3rd edn. (Springer, London, 1995)
15. B. Wang, X. Yu, X. Li, IEEE Trans. Ind. Electr. **55**(11), 4055 (2008)
16. A. Pawlowski, J.L. Guzmn, F. Rodrguez, M. Berenguel, J.E. Normey-Rico, IFAC Proc. Vol. **44**(1), 1779 (2011). 18th IFAC World Congress
17. M.M. Morato, P.R. da Costa Mendes, J.E. Normey-Rico, C. Bordons, IFAC-PapersOnLine **50**(1), 31 (2017). 20th IFAC World Congress
18. Y. Wang, C. Ocampo-Martinez, V. Puig, IET Control Theor. Appl. **10**, 947 (2016)
19. T.Q. Dinh, K.K. Ahn, J. Marco, IEEE Trans. Ind. Electr. **64**(2), 1751 (2017)
20. A. Pawlowski, J.L. Guzmn, F. Rodrguez, M. Berenguel, J. Snchez, in, *IEEE International Symposium on Industrial*. Electronics vol. 2010, pp. 409–414 (2010)
21. X. Wenxia, L. Feijia, L. Shuo, G. Kun, L. Guodong, in *2015 Sixth International Conference on Intelligent Systems Design and Engineering Applications (ISDEA)* (2015), pp. 278–280

Chapter 5
Adaptive Discrete-Time Higher Order Sliding Mode

5.1 Introduction

As noticed in Chaps. 2–4, designing of the sliding mode requires perfect knowledge of the bound on the uncertainties. Generally, this requirement is not fulfilled for the practical systems and the upper bound is either overestimated or underestimated. On one hand, the underestimation of the uncertainties bound deteriorates the invariance property and, on the other hand, the overestimation leads to excessive switching gain. To cope up with the inaccurate estimation problem, adaptive sliding mode is constituted. In this approach, the control gain is adapted dynamically such that the gain is as small as possible but sufficient enough to counter the system uncertainties [1]. Moreover, adaptive sliding mode relaxes requirement of the precise knowledge of the uncertainties bound. Various adaptive sliding mode techniques have been developed for continuous-time systems [2–6].

Although discrete-time first-order sliding mode technique with adaptive gain has been designed earlier in [7], its gain adaptation law causes overestimation of switching gain. These pertinent gaps in the literature form the core motivation of this chapter where an adaptive DHOSM (ADHOSM) is to be devised such that the problem of gain overestimation can be alleviated without the prior knowledge of uncertainty bound.

The major contributions in this chapter are as follows:

1. A higher order reaching law with adaptive switching gain is proposed. The proposed adaptive law allows the gain to decrease within a finite time which helps to avoid the gain overestimation.
2. The proposed technique does not require the knowledge of the bound on the uncertainty and thus, enhances the practical applicability of the controller.

This chapter is organized as follows. The adaptive DHOSM is proposed in Sect. 5.2. The simulation results are presented in Sect. 5.3 followed by conclusion in Sect. 5.4.

© Springer Nature Switzerland AG 2019   71
N. K. Sharma and J. Sivaramakrishnan, *Discrete-Time Higher Order Sliding Mode*,
https://doi.org/10.1007/978-3-030-00172-8_5

5.2 DHOSM with Adaptive Switching Gains

Let us consider a discrete-time representation of a controllable LTI system as given in Chap. 2 by (2.5). However, in this chapter, unlike in Chap. 2, the disturbance bound d^* is not considered to be known. The objective is to design a control input such that DHOSM takes place without the knowledge of disturbance bound d^* and the problem of gain overestimation is avoided as well.

Inspired by the reaching law (2.24), a second-order reaching law with adaptive switching gain is proposed as

$$s(k+1) = k_1 s(k-1) + k_2 s(k) - \varepsilon(k)\mathrm{sgn}[s(k)] - \varepsilon(k)\mathrm{sgn}[s(k-1)] + d(k).$$
(5.1)

The variable switching gain $\varepsilon(k)$ is evaluated from the adaptive laws

$$\varepsilon(k) = \begin{cases} q(k) \text{ if } q(k) > \eta \\ \eta \text{ if } q(k) \leq \eta \end{cases}$$
(5.2)

where

$$q(k) = \begin{cases} \varepsilon(k-1) + \gamma|s(k-1)| & \text{if } \{|s(k)| > |s(k-1)|\}\&\{\mathrm{sgn}[s(k)] = \mathrm{sgn}[s(k-1)]\} \\ \varepsilon(k-1) & \text{if } \{|s(k)| > |s(k-1)|\}\&\{\mathrm{sgn}[s(k)] = -\mathrm{sgn}[s(k-1)]\} \\ \varepsilon(k-1) - \beta|s(k-1)| & \text{if } |s(k)| \leq |s(k-1)| \end{cases}$$
(5.3)

It is to be noted that

$$\varepsilon(0) > \eta.$$
(5.4)

Therein, as given in Chap. 2, the design parameters are chosen as $0 < k_1 < k_2 < 1$ and $k_1 + k_2 < 1$. The constant positive design parameters γ and β are the adaptation rates of the gain and can be tuned as per the required system performance. The parameter η is a positive constant and denotes the minimum value of the gain. It can be tuned according to the disturbance in the system. The DHOSM established through the reaching law (5.1) is called adaptive discrete-time higher order sliding mode (ADHOSM).

It can be noted that the adaptive laws (5.2) with (5.4) imply, $\varepsilon(k) \geq \eta$, $\forall k \geq 0$. Due to the presence of the increasing condition in (5.2), it is necessary to study the upper boundedness of ε which has been carried out in Lemma 5.1.

5.2.1 Proof of Boundedness of the Adaptive Gain

Lemma 5.1 *Given the reaching law (5.1) with the adaptation laws (5.2), $\exists \varepsilon^* \in \mathbb{R}^+$ such that $\varepsilon(k) \leq \varepsilon^* \; \forall k \geq 0$.*

Proof According to the adaptation law (5.2), the switching gain ε decreases for $|s(k)| \leq |s(k-1)|$ and remains $\varepsilon(k) = \eta$. Hence, the verification for the boundedness of ε for these two scenarios is not required. However, ε increases if $|s(k)| > |s(k-1)|$. Thus, in order to ensure boundedness of ε, only the condition $|s(k)| > |s(k-1)|$ is studied in the following cases:

- **Case 1**: If $s(k-1) > 0, s(k) > 0$

In this case, as shown in Fig. 5.1a, the sliding function diverges away from the sliding surface. Let k_{u_0} be any arbitrary finite sampling instant (only used for the analytical purpose) denoting the initial instant of Case 1. The reaching law (5.1) dictates the following dynamics of the sliding function:

$$s(k+1) = k_1 s(k-1) + k_2 s(k) - 2\varepsilon(k) + d(k). \tag{5.5}$$

Further, $s(k) > s(k-1)$ implies

$$s(k+1) < (k_1 + k_2) s(k) - 2\varepsilon(k) + d(k).$$

On considering the maximum value of the disturbance and $k_1 + k_2 < 1$, we obtain

$$s(k+1) < s(k) - 2\varepsilon(k) + d^*.$$

Since ε increases in this case and d^* is a finite constant, there exists a finite sampling instant $k_{u_f} > k_{u_0}$ such that $2\varepsilon(k_{u_f}) \geq d^*$. At this instant, the sliding function is found to be

$$s(k_{u_f} + 1) < s(k_{u_f}). \tag{5.6}$$

It implies that $s(k)$ stops increasing at $k = k_{u_f} + 1$, as shown in Fig. 5.1a. If $s(k_{u_f} + 1) \geq 0$, then $s(k) < s(k-1)$ at $k = k_{u_f} + 1$, and in this situation, $\varepsilon(k)$ decreases according to the third law of (5.2). Otherwise, if $s(k_{u_f} + 1) < 0$, then there exist two possibilities at $k = k_{u_f} + 1$:

1. If $|s(k)| \leq |s(k-1)|$, then $\varepsilon(k)$ decreases as mentioned earlier.
2. If $|s(k)| > |s(k-1)|$, then $s(k) < 0, s(k-1) > 0$, as shown in Fig. 5.1b. The dynamics of s and boundedness for ε for this situation are studied in Case 2.

- **Case 2**: If $s(k) < 0, s(k-1) > 0$

In this case, we have

$$s(k+1) = k_1 |s(k-1)| - k_2 |s(k)| + d(k). \tag{5.7}$$

Fig. 5.1 Possible conditions of adaptive sliding mode

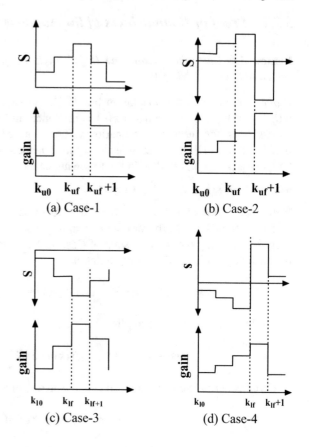

(a) Case-1 (b) Case-2

(c) Case-3 (d) Case-4

It can be noticed in (5.7) that ε has no effect on the dynamics of $s(k)$ in this situation and, following the second condition of the update law (5.2), $\varepsilon(k)$ is kept unchanged at its previous value. Further, k_1 and k_2 are chosen such that the nominal system, obtained on considering $d(k) = 0$ in (5.7), is asymptotically stable at origin. However, as d is bounded, $s(k)$ remains bounded in a band.

It is mentioned in the Case 1, that at a finite time $k = k_{u_f} + 1$, either $\varepsilon(k)$ decreases or the dynamics of $s(k)$ follows Case 2. Now, it is shown that $\varepsilon(k)$ remains unchanged in Case 2 by virtue of the adaptive law (5.2). Hence, on combining the results obtained from Case 1 and Case 2, it can be inferred that $\exists \varepsilon_u^* \in \mathbb{R}^+$ such that $\varepsilon(k) \leq \varepsilon_u^*$ for both the Cases 1 and 2.

- **Case 3**: *If $s(k) < 0, s(k-1) < 0$*

In this case, as shown in Fig. 5.1c, $|s|$ increases with s being negative. To show the boundedness of ε, it is imperative to prove that $|s|$ stops increasing within a finite time.

$$s(k+1) = -k_1|s(k-1)| - k_2|s(k)| + 2\varepsilon(k) + d(k), \tag{5.8}$$

since $s(k), s(k-1) < 0$ in Case 3. Further, the condition $|s(k)| > |s(k-1)|$ implies $-|s(k)| < -|s(k-1)|$. Hence, using the facts $|d(k)| \le d^*$ and $k_1 + k_2 < 1$, (5.8) yields

$$s(k+1) > -k_1|s(k)| - k_2|s(k)| + 2\varepsilon(k) + d(k)$$
$$> -|s(k)| + 2\varepsilon(k) - d^*. \tag{5.9}$$

As ε increases for this case and d^* is finite then, there exist a finite time instant $k_{l_f} > k_{l_0}$ such that $2\varepsilon(k_{l_f}) \ge d^*$. At this instant, from (5.9), we have

$$s(k_{l_f} + 1) > -|s(k_{l_f})|. \tag{5.10}$$

The above condition implies $s(k)$ stops increasing in the negative direction of the sliding surface at $k = k_{l_f} + 1$. If $s(k_{l_f} + 1) \le 0$ then the condition $|s(k)| \le |s(k-1)|$, at $k = k_{l_f} + 1$, is satisfied, which prompts ε to decrease according to the third law of (5.2). Otherwise, if $s(k_{l_f} + 1) > 0$ then there exist two possibilities at $k = k_{l_f} + 1$:

1. If $|s(k)| \le |s(k-1)|$, then $\varepsilon(k)$ decreases as mentioned earlier.
2. If $|s(k)| > |s(k-1)|$, then we have $s(k) < 0, s(k-1) > 0$, as shown in Fig. 5.1d. The dynamics of $s(k)$ and boundedness for ε for this situation are studied subsequently in Case 4.

- **Case 4**: If $s(k) > 0, s(k-1) < 0$

In this case, the following dynamics of s is observed:

$$s(k+1) = -k_1|s(k-1)| + k_2|s(k)| + d(k). \tag{5.11}$$

Similar to the argument made in Case 2, s remains bounded in a band. Further, ε is kept unchanged for this case by the virtue of the adaptive law (5.2).

On combining, Case 3 and Case 4, it can be inferred that there exists an $\varepsilon_l^* \in \mathbb{R}^+$ such that $\varepsilon(k) \le \varepsilon_l^*$ for both these two cases. On observing the boundedness conditions of all the individual possible cases, it can be inferred that

$$\varepsilon(k) \le \varepsilon^*, \quad \forall k \ge 0, \quad \text{where } \varepsilon^* = \max\{\varepsilon_u^*, \varepsilon_l^*\}. \tag{5.12}$$

This completes the proof.

Theorem 5.1 *The sliding function in the reaching law (5.1) with the parameters $0 < k_1 < k_2 < 1$ and $k_1 + k_2 < 1$, is ultimately bounded in a band. Subsequently, the discrete-time higher order sliding mode takes place in a set $\mathcal{M}$, defined in Definition 2.1.*

Proof The ultimate boundedness of the sliding function can be proved by Lyapunov-like stability analysis. Consider $s(k-1) = z_1(k)$, $s(k) = z_2(k)$ and a Lyapunov candidate function

$$V(k) = z_1^2(k) + \alpha z_2^2(k),$$

where α is a positive constant. The first difference of the Lyapunov function can be obtained to be

$$\Delta V(k) = V(k+1) - V(k) \tag{5.13}$$
$$= z_1^2(k+1) + \alpha z_2^2(k+1) - z_1^2(k) - \alpha z_2^2(k) \tag{5.14}$$

On considering

$$\delta(k) = \varepsilon(k)\text{sgn}[s(k-1)] + \varepsilon(k)\text{sgn}[s(k)] - d(k), \tag{5.15}$$

(5.14) can be expanded as

$$= -[1 - \alpha k_1^2]z_1^2(k) - [\alpha - 1 - \alpha k_2^2]z_2^2(k) + \alpha\delta^2(k) + 2\alpha\delta(k)k_1z_1(k) + 2\alpha\delta(k)k_2z_2(k)$$
$$+ 2\alpha k_1 k_2 z_1(k)z_2(k) \tag{5.16}$$

Let $K_a = [1 - \alpha k_1^2]$ and $K_b = [\alpha - 1 - \alpha k_2^2]$. The parameter α is chosen such that $\frac{1}{1-k_1^2} < \alpha < \frac{1}{k_1^2}$. On using the inequality $2z_1(k)z_2(k) \le z_1^2(k) + z_2^2(k)$, the relation (5.16) yields

$$\Delta V(k) \le -K_a z_1^2(k) - K_b z_2^2(k) + \alpha k_1 k_2 [z_1^2(k) + z_2^2(k)] + \alpha\delta^2(k) + 2\alpha\delta(k)k_1z_1(k)$$
$$+ 2\alpha\delta(k)k_2z_2(k) \tag{5.17}$$

K_a and K_b can be separated such that $K_a = K_{a_1} + K_{a_2}$, $K_b = K_{b_1} + K_{b_2}$ and K_{a_1}, $K_{b_1} > \alpha k_1 k_2$.

$$\Delta V(k) \le -K_{a_1} z_1^2(k) - K_{b_1} z_2^2(k) + \alpha k_1 k_2 [z_1^2(k) + z_2^2(k)] - \left[\sqrt{K_{a_2}} z_1(k) - \frac{\alpha\delta(k)k_1}{\sqrt{K_{a_2}}}\right]^2$$
$$- \left[\sqrt{K_{b_2}} z_2(k) - \frac{\alpha\delta(k)k_2}{\sqrt{K_{b_2}}}\right]^2 + \alpha\delta^2(k) + \frac{\alpha^2\delta^2(k)k_1^2}{K_{a_2}} + \frac{\alpha^2\delta^2(k)k_2^2}{K_{b_2}} \tag{5.18}$$

In (5.18), on considering

$$g(k) = \alpha\delta^2(k) + \frac{\alpha^2\delta^2(k)k_1^2}{K_{a_2}} + \frac{\alpha^2\delta^2(k)k_2^2}{K_{b_2}}, \tag{5.19}$$

we get

$$\Delta V(k) \le -\left[K_{a_1} - \alpha k_1 k_2\right] z_1^2(k) - \left[\frac{K_{b_1} - \alpha k_1 k_2}{\alpha}\right] \alpha z_2^2(k) + g(k). \quad (5.20)$$

In (5.2.1), let

$$\phi = \min\left[\left[K_{a_1} - \alpha k_1 k_2\right], \left[\frac{K_{b_1} - \alpha k_1 k_2}{\alpha}\right]\right],$$

where α is chosen such that $0 < \phi < 1$. On using (5.21), (5.20) is deduced to be

$$\Delta V(k) \le -\phi V(k) + g(k).$$

Subsequently, the band of $s(k)$ is found to be

$$|s(k)| \le \sqrt{\frac{g(k)}{\phi \alpha}}.$$

The band of sliding function can be obtained to be

$$|s(k)| \le \beta_d |\delta(k)|, \quad (5.21)$$

where

$$\beta_d = \sqrt{\frac{\left(1 + \frac{\alpha k_1^2}{K_{a_2}} + \frac{\alpha k_2^2}{K_{b_2}}\right)}{\phi}}.$$

On using the value of $\delta(k)$ from (5.15),

$$|s(k)| \le \beta_d(|\varepsilon(k)\mathrm{sgn}[s(k-1)] + \varepsilon(k)\mathrm{sgn}[s(k)] - d(k)|).$$

With $\varepsilon(k) \le \varepsilon^*$ from Lemma (5.1), and the fact $|d(k)| \le d^*$, the following relation holds:

$$|s(k)| \le \beta_d(2\varepsilon^* + d^*).$$

Thus, the sliding function in the reaching law (5.1) is ultimately bounded in a finite band. Therefore, the $\sqrt{\sum_{i=0}^{1}\left(\Delta^i s(k)\right)^2}$ will also be bounded. As the system dynamics are confined to the sliding manifold, the state trajectory will also be ultimately bounded in a set $\mathcal{M}$ and satisfies Definition (2.1). Therefore, the ADHOSM takes place in the set $\mathcal{M}$. This completes the proof.

On using the proposed reaching law (5.1) for the system (2.5), the control input can be synthesized to be

$$u(k) = -(cB)^{-1}\left(cAx(k) - k_1 s(k-1) - k_2 s(k) + \varepsilon_1 T^2 \text{sgn}[s(k-1)] + \varepsilon_2 T^2 \text{sgn}[s(k)]\right).$$
$$(5.22)$$

5.3   Simulation Example

A numerical example of a rectilinear plant is considered for simulation studies. A rectilinear plant is an electromechanical system that represents physical plants with rigid bodies and drives with gear and belt arrangement. It finds applications in processes like automotive, manufacturing machines, CNC machines, and nano-positioning equipments. Consider the model of the rectilinear plant as given in [8] in the form of (2.1) with parameters: $A = \begin{bmatrix} 0 & 1 \\ -1 & -0.1215 \end{bmatrix}$ and $b = \begin{bmatrix} 0 \\ 1 \end{bmatrix}$. The external disturbance $d_x(k) = \begin{bmatrix} 0 & 1.5 \end{bmatrix}^T \times (\sin(k) + \sin(k/3))$ is added to the system. The sliding function is chosen to be $s(k) = \begin{bmatrix} 1 & 2 \end{bmatrix} x(k)$ with the initial conditions $x(0) = \begin{bmatrix} 10 & 10 \end{bmatrix}^T$, $s(-1) = 0$, and $\varepsilon(-1) = 5$. The design parameters k_1 and k_2 are chosen to be 0.03 and 0.4, respectively. The static gains are selected as $\gamma = 0.2$ and $\beta = 0.25$, and $\eta = 0.2$. The proposed ADHOSM is simulated in comparison with the DHOSM presented in [8]. On implementing [8], the switching gain has to be chosen constant at 5. On the other hand, as shown in Fig. 5.2a, the proposed algorithm allows the switching gain to decrease rather than overestimated gain in [8]. As a result, the control, presented in Fig. 5.2b, is lesser compared to [8]. The absolute sum of control input, $\sum_i |u(i)|$, can quantify the total control efforts. This value for the control input using DHOSM control [8] is obtained to be 47.74. On the other hand, on using the proposed ADHOSM, the absolute sum of control input is found out to be 34.03. Furthermore, the sliding function, as shown in Fig. 5.2c, is lesser compared to [8]. Therefore, the proposed technique is better than [8] and utilizes lesser control efforts.

Sometimes, a system comes under the effect of noise. DHOSM control for stochastic systems in the presence of noise is designed in [9]. To demonstrate the performance of the proposed ADHOSM algorithm in the presence of the noise, a noise signal with zero mean and covariance 0.1 is added in the above system. The simulation results are shown in Fig. 5.3 with similar gains selected earlier. It can be noticed that the proposed ADHOSM control provides better performance than the existing DHSOM control even in the presence of noise.

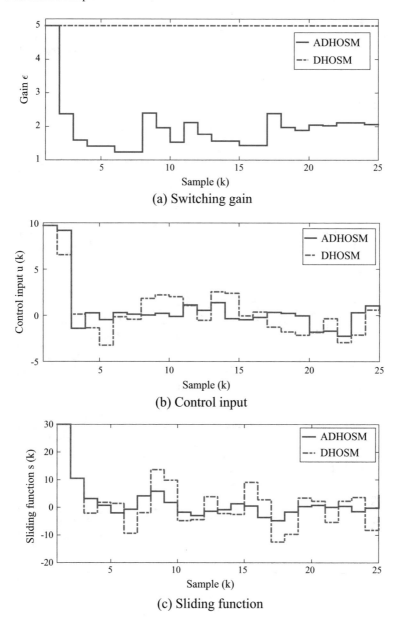

(a) Switching gain

(b) Control input

(c) Sliding function

Fig. 5.2   Simulation results in the presence of disturbance

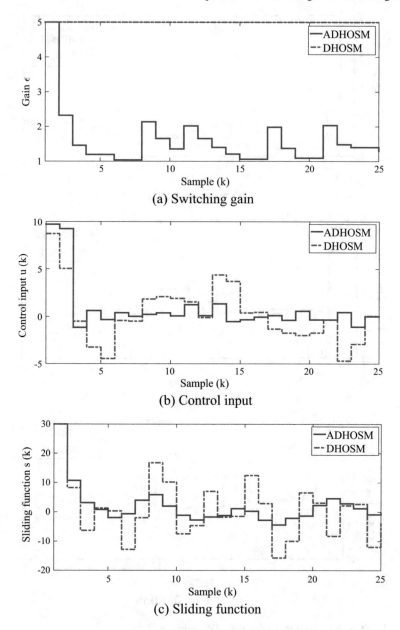

(a) Switching gain

(b) Control input

(c) Sliding function

Fig. 5.3 Simulation results in the presence of disturbance and noise signal

5.4 Conclusion

This chapter proposes an adaptive discrete-time higher order sliding mode controller which allows the establishment of the DHOSM with gain adaptation, without the prior knowledge of the disturbance bound. Further, the adaptive law avoids the over-estimation of the switching gain. Improved performance of the proposed strategy is validated through numerical simulation.

References

1. F. Plestan, Y. Shtessel, V. Brgeault, A. Poznyak, Int. J. Control **83**(9), 1907 (2010)
2. F. Plestan, Y. Shtessel, V. Brgeault, A. Poznyak, Control Eng. Pract. **21**(5), 679 (2013)
3. M. Taleb, F. Plestan, B. Bououlid, Int. J. Robust Nonlinear Control **25**(8), 1201 (2015)
4. Z. Sun, J. Zheng, Z. Man, H. Wang, IEEE Trans. Ind. Electron. **63**(4), 2251 (2016)
5. J. Baek, M. Jin, S. Han, IEEE Trans. Ind. Electron. **63**(6), 3628 (2016)
6. H. Li, P. Shi, D. Yao, IEEE Trans. Automat. Control **62**(4), 1933 (2017)
7. G. Monsees, J.M.A. Scherpen, Int. J. Control **75**(4), 242 (2002)
8. N.K. Sharma, S. Janardhanan, IET Control Theory Appl. **11**, 1098 (2017)
9. N.K. Sharma, S. Singh, S. Janardhanan, D.U. Patil, in *2017 25th Mediterranean Conference on Control and Automation (MED)* (2017), pp. 649–654

Chapter 6
Stochastic Discrete-Time Higher Order Sliding Mode

6.1 Introduction

Many systems, such as communication, supply chain management system, have uncertainties in the probabilistic sense. As a result, these types of systems are better represented by a stochastic model rather than deterministic model. Further, considering a stochastic model for a practical system is more realistic approach in the control system [1]. Although in the earlier chapters, the definitions and design procedures of the discrete-time higher order sliding mode control were proposed considering deterministic model, these definitions do not hold for the stochastic systems. Therefore, there is a need to define and design the discrete-time higher order sliding mode explicitly for the stochastic systems.

With this motivation, the following objectives are contemplated in this chapter:

1. The stochastic version of discrete-time higher order sliding mode (SDHOSM) is defined.
2. A control input is designed for a stochastic system such that stochastic discrete-time higher order sliding mode is established.

The organization of the chapter is as follows. After the introduction in Sects. 6.1, 6.2 proposes the definition of SDHOSM. The problem is formulated in Sect. 6.3. The DHOSM controller is designed in the Sect. 6.4. Section 6.5 contains the simulation results followed by the conclusion in Sect. 6.6.

6.2 Stochastic Discrete-Time Higher Order Sliding Mode

Consider the discrete-time representation of a controllable linear time-invariant (LTI) stochastic system as

$$x(k + 1) = Ax(k) + bu(k) + d_x(k) + \Gamma w(k), \qquad (6.1)$$

© Springer Nature Switzerland AG 2019
N. K. Sharma and J. Sivaramakrishnan, *Discrete-Time Higher Order Sliding Mode*,
https://doi.org/10.1007/978-3-030-00172-8_6

where $x(k) \in \mathbb{R}^n, u(k) \in \mathbb{R}, d_x \in \mathbb{R}^n$ and $w(k) \in \mathbb{R}$ are the state vector, control input, disturbance vector, and process noise with zero mean and covariance matrix V_w. The matrix

$$A = \begin{bmatrix} 0 & 1 & 0 & \cdots & 0 \\ 0 & 0 & 1 & \cdots & 0 \\ \vdots & \vdots & \vdots & \ddots & \vdots \\ -a_1 & -a_2 & -a_3 & \cdots & -a_n \end{bmatrix}, \quad b = \begin{bmatrix} 0 \\ 0 \\ \vdots \\ 1 \end{bmatrix}, \quad d_x = \begin{bmatrix} 0 \\ 0 \\ \vdots \\ d_{x,n} \end{bmatrix}, \quad \text{and} \quad \Gamma = \begin{bmatrix} 0 \\ 0 \\ \vdots \\ 1 \end{bmatrix} \quad \text{are}$$

known and of the appropriate dimensions. All the states are measurable.

Assumption 7 The matched disturbance vector d_x represents the unknown external disturbances of zero mean with the covariance matrix V_d. The disturbance vector d_x is unknown but bounded by known constants [2]. Further, it is assumed that the vector $\Gamma \in Range(b)$. In physical terms, this assumption means that the noise affecting the system is entering through the input channel only.

A sliding function is defined as

$$s(k) = cx(k) = \sum_{i=1}^{r-1} c_i x_i(k) + x_{n-r+1}(k), \tag{6.2}$$

where $s \in \mathbb{R}$ and vector $c = [c_1, c_2, \ldots, c_{n-r}, 1, 0, \ldots, 0] \in \mathbb{R}^n$. It is assumed that the sliding function, s, has a relative degree r with respect to input, u [3, p. 139]. A higher order sliding function is defined as

$$\bar{s}(k) = \bar{c}\xi(k) \tag{6.3}$$

where $\xi(k) = \begin{bmatrix} s_1(k) & s_2(k) & \ldots & s_r(k) \end{bmatrix}^T \in \mathbb{R}^r$, $s_i(k) = s(k - r + i)$ for $i = (1, \ldots, r)$, and vector $\bar{c} = [\bar{c}_1, \bar{c}_2, \bar{c}_3, \ldots, \bar{c}_{r-1}, 1] \in \mathbb{R}^r$. Let us define the discrete-time higher order sliding mode in a stochastic system.

Definition 6.1 (*Stochastic Discrete-time higher order Sliding Mode*) For an uncertain discrete-time system (6.1), $x \in \mathbb{R}^n$ and process noise $w \in \mathbb{R}$, a stochastic discrete-time higher order sliding mode is defined to take place in the band

$$\mathcal{M} = \{x \in \mathbb{R}^n :| \bar{s} |\leq \mu_c\}, \quad \mu_c > 0 \tag{6.4}$$

with a given probability $1 - \delta$, i.e.

$$P\{|\bar{s}(k)| \leq \mu_c\} = 1 - \delta \quad for \quad k > N, \tag{6.5}$$

where $N \in \mathbb{Z}$ is large enough and $0 < \delta \ll 0.5$.

Therefore, the set $\mathcal{M}$ contains the system's state trajectory when the deviation in the higher order sliding function is in the set $[-\mu_c, \mu_c]$ with the probability $1 - \delta$. The parameters μ_c and δ can be designed according to the problem specifications.

6.3 Problem Formulation

The original system given in (6.1) can be transformed to attain a form which is more suitable and easy to design a SDHOSM based control input. The state x in (6.1) can be partitioned into two parts $\bar{x} = [x_1, x_2, ..., x_{n-r}]^T \in \mathbb{R}^{n-r}$ and $\underline{x} = [x_{n-r+1}, x_{n-r+2}, ..., x_n]^T \in \mathbb{R}^r$, and the system (6.1) can be transformed into a new system with the state $z = [\bar{x}, \xi] \in \mathbb{R}^n$. The state z is defined as

$$z(k) = Gx(k), \tag{6.6}$$

where G is a nonsingular state transformation matrix given in (3.9). Thus, the original system (6.1) can be transformed into the system

$$z(k + 1) = \bar{A}z(k) + \bar{b}u(k) + d(k) + \Gamma w(k), \tag{6.7}$$

where $\bar{A} = GAG^{-1}$, $\bar{b} = Gb = [\bar{b}_1 \ \bar{b}_2]^T$ and $Gd_x = d = [d_{\bar{x}} \ d_{\xi}]^T$, $G\Gamma = \Gamma = [\Gamma_{\bar{x}} \ \Gamma_{\xi}]^T$. The system (6.7) can be further partitioned as

$$\begin{bmatrix} \bar{x}(k+1) \\ \xi(k+1) \end{bmatrix} = \begin{bmatrix} \bar{A}_{11} & \bar{A}_{12} \\ \bar{A}_{21} & \bar{A}_{22} \end{bmatrix} \begin{bmatrix} \bar{x}(k) \\ \xi(k) \end{bmatrix} + \begin{bmatrix} 0 \\ \bar{b}_2 \end{bmatrix} u(k) + \begin{bmatrix} 0 \\ d_{\xi}(k) \end{bmatrix} + \begin{bmatrix} 0 \\ \Gamma_{\xi} \end{bmatrix} w(k), \tag{6.8}$$

where

$$\bar{A}_{11} = \begin{bmatrix} 0 & 1 & 0 & \cdots & 0 & 0 \\ 0 & 0 & 1 & \cdots & 0 & 0 \\ \vdots & \vdots & \vdots & \ddots & \vdots & \vdots \\ 0 & 0 & 0 & \cdots & 0 & 1 \\ -c_1 & -c_2 & -c_3 & \cdots & -c_{n-r-1} & -c_{n-r} \end{bmatrix},$$

$$\bar{A}_{12} = \begin{bmatrix} 0 & 0 & \cdots & 0 & 0 \\ 0 & 0 & \cdots & 0 & 0 \\ \vdots & \vdots & \ddots & \vdots & \vdots \\ 0 & 0 & \cdots & 0 & 0 \\ 1 & 0 & \cdots & 0 & 0 \end{bmatrix},$$

$$\bar{A}_{21} = \begin{bmatrix} 0 & 0 & \cdots & 0 \\ 0 & 0 & \cdots & 0 \\ \vdots & \vdots & \ddots & \vdots \\ 0 & 0 & \cdots & 0 \\ -\bar{a}_1 & -\bar{a}_2 & \cdots & -\bar{a}_{n-r} \end{bmatrix},$$

$$\bar{A}_{22} = \begin{bmatrix} 0 & 1 & 0 & \cdots & 0 & 0 \\ 0 & 0 & 1 & \cdots & 0 & 0 \\ \vdots & \vdots & \vdots & \ddots & \vdots & \vdots \\ 0 & 0 & 0 & \cdots & 0 & 1 \\ -\bar{a}_{n-r+1} & -\bar{a}_{n-r+2} & -\bar{a}_{n-r+3} & \cdots & -\bar{a}_{n-1} & -\bar{a}_n \end{bmatrix},$$

$\bar{b}_1 = \mathbf{0}, \bar{b}_2 = \begin{bmatrix} 0 & \cdots & 1 \end{bmatrix}^T \in \mathbb{R}^r, d_{\bar{x}} = \mathbf{0}, d_\xi \in \mathbb{R}^r, \Gamma_{\bar{x}} = \mathbf{0},$ and $\Gamma_\xi \in \mathbb{R}^r$. Let the effect of d_ξ on the higher order sliding function, $\bar{s}$, defined as

$$d(k) = \bar{c} d_\xi(k). \tag{6.9}$$

Since the bounds of the disturbance vector are known, let

$$d_l \leq d(k) \leq d_u. \tag{6.10}$$

The average value of d can be obtained as

$$d_0 = \frac{d_l + d_u}{2}. \tag{6.11}$$

The objective is to find a control input such that the trajectory of the system takes place in a SDHOSM band, defined in (6.4), with the probability as defined in (6.5).

6.4   Stochastic DHOSM Controller

Consider the subsystem of (6.8),

$$\xi(k+1) = \bar{A}_{21}x(k) + \bar{A}_{22}\xi(k) + \bar{b}_2 u(k). \tag{6.12}$$

The proposed stochastic DHOSM control u contains two parts: a direct feedback control u_f and a stochastic control u_{sto}, i.e.

$$u(k) = u_f(k) + u_{sto}(k).$$

As it is assumed that states are measurable, $(n - r + 1)$ states can be used to design the direct feedback control part as

$$u_f(k) = -(\bar{c}\bar{b}_2)^{-1}\bar{c}\bar{A}_{21}\bar{x}(k) = \sum_{i=1}^{n-r} \bar{a}_i x_i(k). \tag{6.13}$$

On utilizing the feedback control input (6.13), the subsystem (6.12) reduces to

$$\xi(k+1) = \bar{A}_{22}\xi(k) + \bar{b}_2 u_{sto}(k).\tag{6.14}$$

Now, a control input is to be designed for (6.14) such that the higher order sliding function enters into the band μ_c in finite time.

6.4.1 Convergence Analysis

The sufficient condition to ensure the ultimate boundedness of $\bar{s}$ in a band can be provided if the reaching law for $\bar{s}$ is chosen such that

$$|\bar{s}(k+1)| < \theta|\bar{s}(k)|,\tag{6.15}$$

where $0 < \theta < 1$ regulates the speed of the convergence of $\bar{s}$. The condition (6.15) is known as the approaching condition for the sliding function and ensures the finite time reaching phase. The finite time convergence of (6.15) is proved in [4].

Lemma 6.1 ([4]) *Let g be a Gaussian random variable with zero mean and variance σ_g^2, and W_g is the solution of the equation*

$$P\{|g| \le W_g\} = \int_{-W_g}^{W_g} \frac{1}{\sigma\sqrt{2\pi}} e^{-z^2/2\sigma^2} dz = 1 - \varepsilon.$$

Let $g_m = m + g$. Then the solutions of the following equation concerning m

$$P\{|g_m| \le L_m\} = \int_{-L_m}^{L_m} \frac{1}{\sigma\sqrt{2\pi}} e^{\frac{-(z-m)^2}{2\sigma^2}} dz = 1 - \varepsilon \tag{6.16}$$

have following properties:

(i) if $L_m > W_g$, then (6.16) has only two solutions.
(ii) if $L_m > W_g$ and $0 < \varepsilon < 0.5$, then the solutions of (6.16) are bounded by L_m, i.e. $|m| \le L_m$.

6.4.2 Controller Design

The aim is to find a control input such that the objective (6.5) is achieved. From (6.3), we get

$$\bar{s}(k+1) = \bar{c}\xi(k+1).\tag{6.17}$$

Let the higher order sliding function be partitioned into deterministic and probabilistic parts as

$$\bar{s}(k+1) = m(k+1) + g(k+1), \tag{6.18}$$

where the deterministic part is

$$m(k+1) = \bar{c}\bar{A}_{22}\xi(k) + \bar{c}\bar{b}_2 u_{sto}(k) + d_0, \tag{6.19}$$

and the probabilistic part is

$$g(k+1) = d(k) - d_0 + \bar{c}\Gamma_\xi w(k). \tag{6.20}$$

It can be noticed that $g(k+1)$ is Gaussian with zero mean and variance $\sigma_c^2 = \bar{c}(V_d + \Gamma V_w \Gamma^T)\bar{c}^T$.

Theorem 6.1 *The objective,*

$$P\{|\bar{s}(k)| \leq \mu_c\} = 1 - \delta \quad for \quad k > N,$$

for the system (6.14) can be achieved by the use of control input

$$u_{sto}(k) = -\left(\bar{c}\bar{b}_2\right)^{-1}(\bar{c}A_{22}\xi(k) + d_0).$$

Proof On the basis of the value of the higher order sliding function, $\bar{s}$, the following three cases are possible:
   Case 1: If $\bar{s}(k) > \theta^{-1}W_c(k)$
On considering the above case,

$$P\{-\theta\bar{s}(k) < \bar{s}(k+1) < \theta\bar{s}(k)\} \geq 1 - \varepsilon. \tag{6.21}$$

On using (6.18) in (6.21), we get

$$P\{-\theta\bar{s}(k) < m(k+1) + g(k+1) < \theta\bar{s}(k)\} \geq 1 - \varepsilon. \tag{6.22}$$

The necessary and sufficient condition for the existence of solutions of (6.22) is

$$\bar{s}(k) > \theta^{-1}W_c, \tag{6.23}$$

where W_c can be obtained from the relation

$$\int_{-W_c}^{W_c} \frac{1}{\sigma_c\sqrt{2\pi}} e^{\frac{-z^2}{2\sigma_c^2}} dz = 1 - \varepsilon.$$

On solving the above equation, we get

$$W_c = \sqrt{2}\sigma_c \ (\mathbf{erf})^{-1}(1-\varepsilon), \tag{6.24}$$

where **erf** represents the error function defined as

$$\mathbf{erf}(y) = \frac{2}{\sqrt{\pi}} \int_0^y e^{-t^2} dt. \tag{6.25}$$

On considering Lemma 6.1 and the condition (6.23), the control input can be obtained to be

$$u_{sto}(k) = u_{sto}^+(k) \in (\bar{c}\bar{b}_2)^{-1}\left(-(\bar{c}A_{22}\xi(k) + d_0) + [m_1^+(k), m_2^+(k)]\right). \tag{6.26}$$

Here $[\beta_1, \beta_2]$ denotes the closed interval with limit points β_1 and β_2. The $m_1^+(k)$ and $m_2^+(k)$ are the only two solutions of

$$\int_{-\theta\bar{s}(k)}^{-\theta\bar{s}(k)} \frac{1}{\sigma_c\sqrt{2\pi}} e^{\frac{-(z-m)^2}{2\sigma_c^2}} dz = 1 - \varepsilon,$$

such that $m_1^+(k) < m_2^+(k)$ is satisfied.

Case 2: If $\bar{s}(k) < -\theta^{-1}W_c(k)$

On considering the above case, the relation (6.15) yields

$$P\{\theta\bar{s}(k) < m(k+1) + g(k+1) < -\theta\bar{s}(k)\} \geq 1 - \varepsilon. \tag{6.27}$$

The necessary and sufficient condition for the existence of the solutions of (6.27) is

$$\bar{s}(k) < -\theta^{-1}W_c. \tag{6.28}$$

Similar to case 1, in this case, the control input can be obtained to be

$$u_{sto}(k) = u_{sto}^-(k) \in (\bar{c}\bar{b}_2)^{-1}(-(\bar{c}A_{22}\xi(k) + d_0) + [m_1^-(k), m_2^-(k)]), \tag{6.29}$$

where $m_1^-(k)$ and $m_2^-(k)$ are the only two solutions of

$$\int_{\theta\bar{s}(k)}^{-\theta\bar{s}(k)} \frac{1}{\sigma_c\sqrt{2\pi}} e^{\frac{-(z-m)^2}{2\sigma_c^2}} dz = 1 - \varepsilon,$$

such that $m_1^-(k) < m_2^-(k)$.

Case 3: If $-\theta^{-1}W_c(k) \leq \bar{s}(k) \leq \theta^{-1}W_c(k)$, i.e. $|\bar{s}(k)| \leq \theta^{-1}W_c$.

This case yields $m_1 = m_2 = 0$. Hence, the control input is equal to equivalent control, i.e.

$$u_{sto}(k) = u_{eq}(k) = -(\bar{c}\bar{b}_2)^{-1}(\bar{c}A_{22}\xi(k) + d_0). \tag{6.30}$$

Therefore, on utilizing the control inputs (6.26), (6.29) and (6.30), the higher order sliding function, $\bar{s}$, will be ultimately bounded in a band with a probability $1 - \delta$. With help of Lemma 6.1, it is straight forward that (6.30) would satisfy the range of $u_{sto}(k)$. Thus, the equivalent control input (6.30) will be appropriate to use as the control input to achieve the objective (6.5) [4]. This completes the proof.

Thus, the combined control input, on using (6.13) and (6.30), can be obtained to be

$$u(k) = \sum_{i=1}^{n-r} \bar{a}_i x_i(k) - (\bar{c}\bar{b}_2)^{-1}(\bar{c}A_{22}\xi(k) + d_0). \tag{6.31}$$

Let $(\bar{c}\bar{b}_2)^{-1}\bar{c}A_{22} = \begin{bmatrix} f_1 & f_2 & \cdots & f_r \end{bmatrix} \in \mathbf{R}^r$ for the simplicity of representation. The combined control is obtained to be of the form,

$$u(k) = \sum_{i=1}^{n-r} \bar{a}_i x_i(k) - F\xi(k) - (\bar{c}\bar{b}_2)^{-1}d_0). \tag{6.32}$$

6.4.3 Stability and Covariance Analysis

On using the combined control input (6.4.2) in the system (6.7), the resulting closed-loop system is obtained to be

$$\bar{x}(k+1) = \bar{A}_{11}\bar{x}(k) + \bar{A}_{12}\xi(k), \tag{6.33}$$
$$\xi(k+1) = A_c\xi(k) + \delta_\xi(k) + \Gamma_\xi w(k), \tag{6.34}$$

where

$$A_c = \begin{bmatrix} 0 & 1 & \cdots & 0 & 0 \\ 0 & 0 & \cdots & 1 & 0 \\ \vdots & \vdots & \ddots & \vdots & \vdots \\ 0 & 0 & \cdots & 0 & 1 \\ -\bar{a}_{n-r+1} + f_1 & -\bar{a}_{n-r+1} + f_2 & \cdots & -\bar{a}_{n-1} + f_{r-1} & -\bar{a}_n + f_r \end{bmatrix}, \tag{6.35}$$

$$\delta_\xi = d_\xi - \bar{b}_2 d_0 \tag{6.36}$$

The covariance of state $\bar{x}$ and ξ in the closed loop can be obtained as follows. Consider (6.34),

$$\xi(k+1) = A_c\xi(k) + \delta_\xi(k) + \Gamma_\xi w(k).$$

At nth sample, ξ is obtained to be

$$\xi(n) = A_c^n \xi(0) + L_1 + L_2, \tag{6.37}$$

where

$$L_1 = \left(A_c^{n-1} \delta_\xi(0) + \cdots + A_c \delta_\xi(n-2) + \delta_\xi(n-1) \right)$$
$$L_2 = \left(A_c^{n-1} \Gamma w(0) + \cdots + A_c \Gamma w(n-2) + \Gamma w(n-1) \right)$$

Since the disturbance is bounded and its bounds are known, there exists a known δ_u such that $||\delta_\xi|| \leq \delta_u$. The norm of L_1,

$$||L_1|| \leq || \left(A_c^{n-1} + \cdots + A_c + I \right) ||\delta_u,$$
$$\leq ||[I - A_c^n][I - A_c]^{-1}||\delta_u. \tag{6.38}$$

The vector $\bar{c}$ is chosen such that the spectral radius of A_c is less than one. Therefore, at steady state $n \to \infty$, ξ is bounded in a band given by

$$||\xi|| \leq \left\| [I - A_c]^{-1} \right\| \delta_u = \beta.$$

The nature of L_2 is probabilistic and its covariance can be obtained to be,

$$V_{L2} = \sum_{i=0}^{\infty} (A_c^i \Gamma) V_w (A_c^i \Gamma)^T. \tag{6.39}$$

Then, V_{L2} is the solution of

$$V_{L2} = \Gamma V_w \Gamma^T + A_c V_{L2} A_c^T. \tag{6.40}$$

The above equation is in the form of discrete-time Lyapunov equation and can be solved using the numerical methods to obtain the covariance V_{L2} [5]. Thus, ξ is stable in a band with the covariance of V_{L2}.

Further, it can be noticed that state $\bar{x}$ in (6.33) depends upon ξ. The values of vector c are chosen such that the spectral radius of $\bar{A}_{11}$ is less than one. Therefore, the covariance of $\bar{x}$ can be obtained with the same procedure as given above. Consider (6.33)

$$\bar{x}(k+1) = \bar{A}_{11}\bar{x}(k) + \bar{A}_{12}\xi(k) \tag{6.41}$$

At nth sample, $\bar{x}$ is obtained to be

$$\bar{x}(n) = \bar{A}_{11}^n \bar{x}(0) + \bar{A}_{12}\xi(0) + \bar{A}_{12}\bar{A}_{11}\xi(1) + \bar{A}_{12}\bar{A}_{11}^2\xi(2) + \cdots + +\bar{A}_{12}\bar{A}_{11}^{n-1}\xi(n-1), \tag{6.42}$$

$$= \bar{A}_{11}^n \bar{x}(0) + \bar{A}_{12} \sum_{i=0}^{n-1} \bar{A}_{11}^i \xi(i). \tag{6.43}$$

Thus,

Table 6.1 Parameters value

Parameter	Value
V_w	0.1
θ	0.9
ε	0.3

$$\|\bar{x}(n)\| \leq \left\|\bar{A}_{11}^n\right\| \|\bar{x}(0)\| + \left\|\bar{A}_{12}\right\| \left\|\sum_{i=1}^{n-1} \bar{A}_{11}^i\right\| \|\xi\|, \tag{6.44}$$

$$\|\bar{x}(n)\| \leq \left\|\bar{A}_{11}^n\right\| \|\bar{x}(0)\| + \left\|\bar{A}_{12}\right\| \left\|\left[I - \bar{A}_{11}^n\right]\left[I - \bar{A}_{11}\right]^{-1}\right\| \|\xi\|. \tag{6.45}$$

As parameter c is chosen such that spectral radius of $\bar{A}_{11}$ is less than one, at steady state

$$\|\bar{x}\| \leq \left\|\bar{A}_{12}\right\| \left\|\left[I - \bar{A}_{11}\right]^{-1}\right\| \beta. \tag{6.46}$$

The covariance of $\bar{x}$ in the steady state can be obtained to be

$$V_{\bar{x}} = \sum_{i=0}^{\infty} \left(\bar{A}_{12}\bar{A}_{11}^i\right) V_{L2} \left(\bar{A}_{12}\bar{A}_{11}^i\right)^T \tag{6.47}$$

Then, $V_{\bar{x}}$ is the solution of

$$V_{\bar{x}} = \bar{A}_{12} V_{L2} \bar{A}_{12}^T + \bar{A}_{11} V_{\bar{x}} \bar{A}_{11}^T. \tag{6.48}$$

Equation (6.48) can be solved using the numerical methods [5]. Therefore, $\bar{x}$ is stable in a band, given in (6.46), with the covariance $V_{\bar{x}}$.

6.5 Simulation Results

To verify the performance of the proposed controller, a discrete-time system (6.1) is simulated with parameters (Table 6.1)

$$A = \begin{bmatrix} 0 & 1 & 0 & 0 & 0 & 0 \\ 0 & 0 & 1 & 0 & 0 & 0 \\ 0 & 0 & 0 & 1 & 0 & 0 \\ 0 & 0 & 0 & 0 & 1 & 0 \\ 0 & 0 & 0 & 0 & 0 & 1 \\ -0.7 & 0.9 & -1.1 & 1.2 & -1.14 & 1.1 \end{bmatrix}, \bar{b} = \begin{bmatrix} 0 \\ 0 \\ 0 \\ 0 \\ 0 \\ 1 \end{bmatrix}, d(k) = \begin{bmatrix} 0 \\ 0 \\ 0 \\ 0 \\ 0 \\ \left(\sin(\frac{k}{2}) + \sin(\frac{k}{100})\right) \end{bmatrix}.$$

The initial value of the state vector is set to be $x(0) = \begin{bmatrix} 10 & 10 & 10 & 10 & 10 & 10 \end{bmatrix}^T$.

The simulation results are presented in Figs. 6.1, 6.2 and 6.3. Figure 6.1 shows that the higher order sliding function is stabilized in a band with $W_c = 0.3277$, obtained in Sect. 6.4. Since the θ is chosen to be 0.9, it can be noticed that the convergence speed

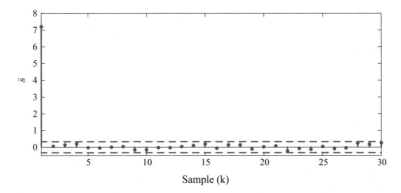

Fig. 6.1 Higher order sliding function on using SDHOSM control

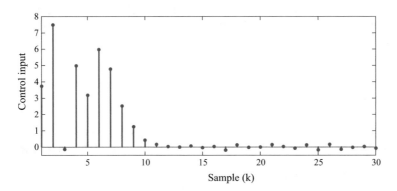

Fig. 6.2 Control input $u(k)$ on using SDHOSM algorithm

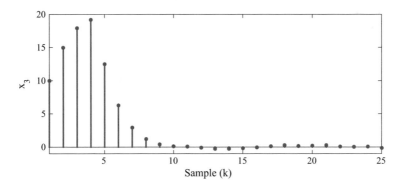

Fig. 6.3 States of the system on using SDHOSM control

of sliding function is very high. The lower value of θ will decrease the convergence speed accordingly. The control input, shown in Fig. 6.2, also becomes stable once the higher order sliding function is stable in a band. It is also seen in the Fig. 6.3

that the state trajectory of state x_3 converges in the steady state. The other states are also stable similar to x_3. Therefore, the proposed definition of the stochastic discrete higher order sliding mode is validated and the proposed control proved to stabilize the uncertain stochastic system.

6.6   Conclusion

In this chapter, a definition of stochastic discrete higher order sliding mode is proposed. A control law is proposed for an uncertain stochastic system such that the defined SDHOSM takes place in a band. The bandwidth of the SDHOSM band depends upon the control parameters and can be tuned accordingly. The proposed control algorithm is simulated and it was seen that the simulation result validates the proposed definition of SDHOSM. The designed control input successfully stabilizes the uncertain stochastic system.

References

1. B. Zhao, Y. Peng, F. Deng, IET Control Theory Appl. **11**(16), 2910 (2017)
2. S. Singh, S. Janardhanan, in *2015 International Workshop on Recent Advances in Sliding Modes (RASM)* (2015), pp. 1–6
3. A. Isidori, *Nonlinear Control Systems*, 3rd edn. (Springer, London, 1995)
4. F. Zheng, M. Cheng, W.B. Gao, Syst. Control Lett. **22**(3), 209 (1994)
5. A.Y. Barraud, in *1977 IEEE Conference on Decision and Control including the 16th Symposium on Adaptive Processes and A Special Symposium on Fuzzy Set Theory and Applications* (1977), pp. 420–423

Index

© Springer Nature Switzerland AG 2019
N. K. Sharma and J. Sivaramakrishnan, *Discrete-Time Higher Order Sliding Mode*,
https://doi.org/10.1007/978-3-030-00172-8

Printed in the United States
By Bookmasters